METAL MATRIX COMPOSITES IN INDUSTRY

An Introduction and a Survey

DISCLAIMER

This document aims only to provide a description of industrial activity in a specific area of technology. It is not a handbook, nor is it a buyer's guide. Its contents should not be used in design and should not be used to compare companies or commercial products.

The text and the accompanying database were written free of commercial interest, and have been kept free of any advertisement or other commercial bias. Data included here, despite having been collected and reported with care, are certainly not complete nor flaw-free; hence, they should not be be used without direct confirmation from suppliers of their validity and measurement method.

Neither the Swiss Federal Institute of Technology in Lausanne (EPFL), nor the authors, nor any person associated with this effort including the Publisher, will assume any responsibility for usage by the reader of the information herein.

Metal Matrix Composites in Industry: An Introduction and a Survey

by

ALEXANDER EVANS[1]

CHRISTOPHER SAN MARCHI[2]

ANDREAS MORTENSEN

Laboratory for Mechanical Metallurgy, Institute of Materials
Swiss Federal Institute of Technology in Lausanne (EPFL)
CH-1015, Lausanne, Switzerland

[1] *(Now at Manchester Materials Science Centre, Grosvenor Street, Manchester, M1 7HS, United Kingdom)*
[2] *(Now at Sandia National Laboratories, PO Box 969, MS-9402, Livermore CA 94551-0969, USA)*

Incorporating a database by:

Christopher San Marchi, Alexander Evans, Cedric Gaillard, Jean-François Despois, Bastien Ecabert, and Andreas Mortensen

SPRINGER SCIENCE+BUSINESS MEDIA, LLC

Library of Congress Cataloging-in-Publication Data

Evans, Alexander
Metal matrix composites in industry: an introduction / by Alexander Evans, Christopher San Marchi, Andreas Mortensen
p. cm.
Incorporating a database by Christopher San Marchi ... et al.

Additional material to this book can be downloaded from http://extras.springer.com.

ISBN 978-1-4020-7521-6 ISBN 978-1-4615-0405-4 (eBook)

DOI 10.1007/978-1-4615-0405-4

1. Metallic composites. I. San Marchi, Christpoher, II, Mortensen A. (Andreas) III. Title

TA481.E93 2003
620.1'6—dc21

2003054988

Printed on acid-free paper.

CONTENTS

PREFACE

Metal matrix composites (MMC) have been the subject of scientific investigation and applied research for about three decades, but these advanced materials have been introduced only gradually in particular engineering components, such as space shuttle and satellite components, jet fighter aircraft parts, combustion engine components, brake rotors or heat sinks for electronic systems. Their greatest asset is the combination of properties of a metal matrix with those of a reinforcement, usually ceramic, providing a potential for tailoring material properties to meet specific and challenging requirements. There is a great variety of material combinations (ingredients, volume fractions and shape, processing) giving rise to a wide spectrum of types of MMC, each with a specific property profile. On the other hand this extraordinary variety complicates the know-how transfer to industry and hinders the acceptance by engineers developing new products. This presentation will help to orient engineers and potential users.

A European network *"Assessment of Metal Matrix Composites for Innovations"* [1-3] was formed to improve the dissemination of scientific results on MMC and to increase market acceptance by providing assessed information of the outstanding potential of MMC with their specific limitations. This thematic network, bearing the acronym "MMC-Assess", was funded to a great extent by the European Community during the period October 1998 until December 2001. Twenty-one partners from industry (6), research institutions (8), universities (6) and the German Materials Society co-operated with further collaborating researchers forming an EU network comprising more than 100 MMC-experts with links to Eastern Europe, USA [4] and Japan. This overview of the industrial landscape on MMC is one of the products of these networking activities.

In setting up the network, working groups were established according to a matrix formed by the categories of MMC and the topics of processing, characterisation, properties and application, which are common to all types of MMC. Table 1 outlines the working group matrix indicating the chairing organisations. The topical working groups elaborated the state-of-the-art assessments in their subjects, whereas the four material-related working groups collected representative information describing selected MMC, which was assessed during working group meetings and plenary sessions of the partners. The strengths, weaknesses, opportunities and threats (SWOT) related to the application of MMC have been analysed and discussed. Additionally, workshops and seminars on specific topics have been held at universities and industrial companies. A symposium on

Table 1: Working group matrix for the MMC-Assess network

Particle Reinforced Metals (PRM)	Short Fibre Reinf. Metals (SFRM)	Continuous Fibre Reinf. Metals (CFRM)	Mono-Filament Reinf. Metals (MFRM)
Dept.Mat.Sci.& Met., U. Cambridge, UK	GKSS-Res.Centre Geesthacht (DGM), D	EADS-Germany, D	DLR-Germ.Aerospace Centre/Cologne, D

Ingredients & Pre-processing	Institute of Composite Materials and Surface Technology, TU.Chemnitz, (DGM), D
Primary Processing & Recycling	Swiss Federal Institute of Technology in Lausanne (EPFL), CH
Casting, Forming, Machining, Joining	CSM Materialteknik, S
Microstructure & Heat Treatment	Inst.Nat.Science Appliquee/Metall. & Mat. Phys. Group (GEMPPM), INSA Lyon, F
Interface Characteristics	Eidgenössische Materialprüfungs- u. Forschungsanstalt Thun (EMPA Thun), CH
Modelling & Simulation	Depto.de Ciencia de Materiales, Universidad Politécnica de Madrid (UPM), E
Mechanical Properties	Centre for Materials Measurement & Techn., National Physics Laboratory (NPL), UK
Physical Properties	Austrian Res.Centres Seibersdorf (ARCS), A
Surface, Corrosion & Wear	Inst.Ceramic Mat., TU.Freiberg (DGM), D
Quality Control & Non-destructive Testing	Department of Metallurgy & Materials Eng., Katholieke Universities Leuven (KUL), B
Applications	INTOSPACE, D
Air & Rail Applications	Aerospatiale (EADS-CCR), F
Automotive Applications	Centro Ricerche Fiat (CRF), IT
Electrotechnical & Functional Applications	EA-Technology (Capenhurst-Technology), UK
Machinery, Sports, Dual Use	MMC Committee, Inst. of Materials, UK
SWOT Assessment	INTOSPACE, D

Abbreviations: DGM (German Society for Materials Science), EADS (European Aeronautic, Defence & Space company)

Table 2: Menu of the homepage "http://mmc-assess.tuwien.ac.at/"

"who we are"	Information about MMC-Assess and contact addresses
"what are MMC's"	MMC terminology
"applications"	Existing and potential applications
"material data"	Data & information on selected MMC
"supplier"	Description of producers of MMC or components (contacts)
"literature"	Selected publications correlated to topics
"conferences"	Permanently updated MMC events
"mmc-links"	Structured links to MMC on the Web
"news"	MMC-Assess activities: meetings, seminars, workshops
"member"	Network documents (restricted access for network partners)
"Assess-publications"	Publications of MMC-ASSESS (for downloading)
"jobs"	Job exchange platform for MMC-experts

X-ray computed tomography was organized together with INSA-Lyon [5] on the topic of characterization and non-destructive testing of MMC. The state-of-the-art on MMCs was presented at different conferences via lectures and by an exhibition of prototypes and industrial products made of MMC.

A web page was designed [6], which focuses on MMC-relevant topics, accessible under the URL "http://mmc-assess.tuwien.ac.at/". It serves as the main platform for information exchange within the Network and the research community, providing a worldwide accessible publishing medium. It enhances the dissemination of relevant engineering information on MMC and registers the user needs. The most recent version comprises header titles that are listed in Table 2. The homepage is permanently expanded and webspace is available for the presentation of interested companies or research institutions active in the field of metal matrix composites.

The data bank on MMC producers, suppliers and users that was designed and kept at EPFL [7] over the course of the network formed the basis for the overview of the industrial situation of MMCs that is presented here. A.Mortensen contributed essentially to the development of MMC since almost two decades. He is aware of the international state-of-the-art playing himself an important role in MMC research not only in Europe but as well in the USA and in cooperation with Japan. I have worked with him in many years on this subject, and share with him the vision that these materials still represent a largely untapped technology for many applications. The information provided in this industry-focused overview on metal matrix composites should be useful to many in the "real" world of engineering.

Prof. H.Peter Degischer,
Co-ordinator of the "MMC-ASSESS" network
Institute of Materials Science and Testing, Vienna University of Technology,
Karlsplatz 13/E308, A-1040 Vienna
http.//www.tuwien.ac.at/E308

Vienna, October 2002

References:

[1] "A road to European cooperation- thematic networks activity, IMT Programme 1994-98", EU Commission DGXII, EU 18197, Brussels 1999, 144-145 (available via http://www.cordis.lu/).

[2] "Assessment of Metal Matrix Composites for Innovations" (MMC-ASSESS)- EU Thematic Network Program, BET2-261, BRRT-CT98-5061, co-ordinated by H.P.Degischer, TU-Vienna/EU-DGXII-Brussels, 1998; (abstract via internet http://www.cordis.lu/).

[3] H.P. Degischer, P. Prader, C.S. San Marchi: *Assessment of metal matrix composites for innovations: intermediate report of an EU Thematic Network,* Comp.A, 32 (2001), 1161-1166.

[4] "Aluminium Matrix Composites Consortium", USA, http://www.almmc.com/ (opened 1998, continuously up-dated).

[5] J.Baruchel et al., *X-ray Tomography in Materials Science,* Hermes Science Publications, Paris 2000.

[6] P.Prader et al., Homepage: *http://mmc-assess.tuwien.ac.at/,* Vienna University of Technology, (opened 1999, continuously up-dated).

[7] A. Mortensen et al., *Industrial landscape on MMC,* filemaker data bank accessible via http://dmxwww.epfl.ch/lmm (opened 1999, updated 2001).

ACKNOWLEDGMENTS

Funding of this work as part of the EU Thematic Network BET2-621 "MMC-ASSESS" by the Swiss Federal Government through the Swiss Federal Office for Education and Science (BBW/OFES, contract no. BRRT-CT98-5061), is gratefully acknowledged. Thanks are expressed to all the Network partners and participants, especially the coordinating topic leaders directly involved in this effort: H. Persson (CSM Materialteknik, Sweden), C. Fink (Intospace, Belgium) P. Hayes (EA-Technology, Great Britain), and of course also the initiator and leader of this network, Professor H.P. Degischer of TU-Wien, Austria.

This book was assembled and formatted according to Publisher's instructions by Ms. Nathalie Zanetta, whom we thank for the care and dedication she put in this task. We also thank Mr Willy Dufour, who brought the computer database to final form before inclusion in the book.

We gratefully acknowledge the contribution of Jean-François Despois, Bastien Ecabert and Cédric Gaillard, our co-authors on the database, in researching the MMC industry on the internet over the course of the project; also, we thank Dr. Cédric Gaillard for assembling the tables in Chapters 2 and 3.

Last but not least, we wish to express our gratitude to all of our colleagues in industry who so kindly provided the information and documentation necessary to support the information in this survey, and to our families for their patience and support.

Chapter 1

INTRODUCTION

1.1 Definition of Metal Matrix Composites

A composite material is a mixture of two or more separate phases, which have been intimately bonded together at a scale that is sufficiently fine that the result can be considered a material with properties of its own.

A metal matrix composite (MMC) is a composite that contains a continuous metallic matrix, and a reinforcement that represents at least a few percent of the material by volume. Essential elements of this definition are:

(i) it refers to a *material*, the mesoscopically averaged properties of which can be used in design of components of various shapes and sizes, as long as the distribution of the constituents is fine on the scale of the component. As such, a composite material is distinct from layered or laminated components, or from macroscopic multimaterial structures;

(ii) a metal matrix composite is the result of a physical combination of ingredient materials which remain distinct throughout the processing history of the material. This is different from conventional alloys, including eutectics or alloys containing precipitates or segregated inclusions, in which the various phases present are all formed from a homogeneous melt during solidification and/or heat-treatment.

This definition thus excludes directionally solidified eutectics, as well as Al-Si alloys or cast iron, all of which are sometimes called metal matrix composites. This definition still includes many *in-situ* composite materials: any material that contains a significant fraction of two phases that remain distinct throughout its processing history, qualifies as a composite (even if the chemical composition of the second phase changes).

Figure 1.1 provides an example of a metal matrix composite (a ceramic particle reinforced aluminium with about 50% ceramic), and of a well-established metal matrix composite production process (the pressure-infiltration process).

B4C
Al2O3

(a)

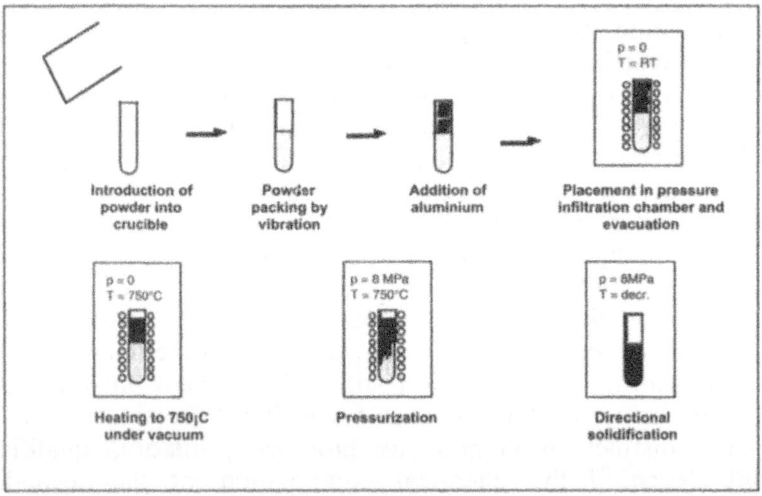
p = 0
T = RT
Introduction of powder into crucible
Powder packing by vibration
Addition of aluminium
Placement in pressure infiltration chamber and evacuation
p = 0
T = 750°C
p = 8 MPa
T = 750°C
p = 8MPa
T = decr.
Heating to 750¡C under vacuum
Pressurization
Directional solidification

(b)

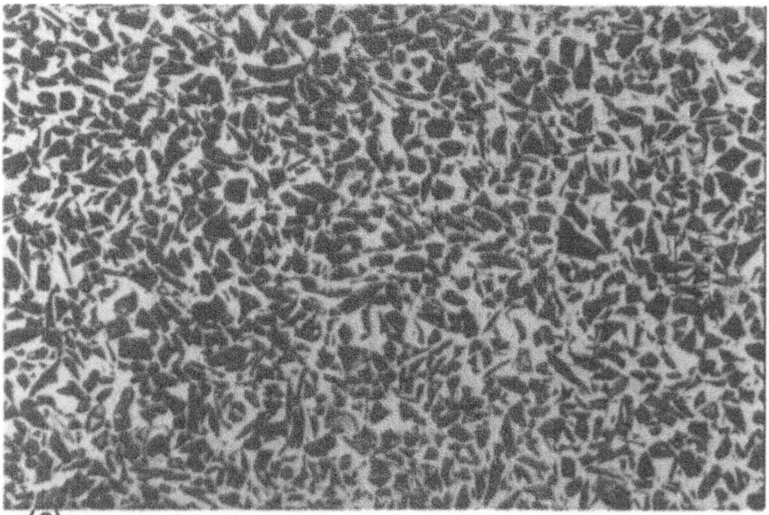
(c)

Figure 1.1 - A metal matrix composite *(a)* and its production sequence *(b)*: *(a)* ingredient materials, namely an aluminium ingot (diameter 36 mm) and ceramic particulate reinforcements (alumina or boron carbide). These are placed one over the other in an alumina crucible (also shown in *(a)*), then heated under vacuum in a pressure vessel *(b)*. Once the metal is molten, the chamber is pressurized; the metal is then driven to flow into the powder bed underneath it. After cooling and matrix solidification, a composite is produced, combining the metal with the reinforcement that was placed underneath in the crucible. An example is given in *(c)*, which shows the microstructure of an Al/Al_2O_3 composite, viewed under the reflective optical microscope after metallographic polishing: the aluminium matrix is white, the alumina particles are darker (from current research at EPFL).

1.2 Metal Matrix Composites as Composites

Composite materials are now very familiar: “fiberglass” (somewhat more properly called a glass fibre reinforced polymer) is for example a ubiquitous element of our daily life. Yet, composites are also high-tech materials, recognized by the broader public for the edge they provide in various sports, such as tennis, golf, and sailing. Engineers are now also familiar with composites as important materials that constitute an increasing proportion of the modern airplane and automobile. Most composites utilized in the sporting, aerospace and automotive industries consist of a reinforcing phase, such as glass or carbon fibres, with a polymeric “glue” or matrix. Polymers are a good choice for matrix materials as they are inexpensive, light, and easy to handle and process when making the composite.

Compared with polymer matrix composites, metal matrix composites are significantly more difficult to produce. The expense and difficulty of producing metal matrix composites, however, can be offset by the property advantages of the metal matrix over a polymer, since these are transposed to the composite. Metals uniquely combine strength and toughness and can maintain these properties at elevated temperatures. Unlike polymers, metals do not absorb humidity, are not degraded by radiation, do not outgas in space, and generally do not burn. Metal matrix composites also conduct heat and electricity, which is an important attribute in some applications. In many cases, metal matrix composites can be welded or easily fastened to other metallic components. Polymer and metal matrix composites thus have rather different characteristics, and thus they also have different fields of application.

1.3 Metal Matrix Composites as Metals

Conventional alloys often achieve remarkably high performance: highly drawn pearlitic steel wire is one of the strongest materials available, and tool steels provide the basis for many of the "hard" tools used in shaping and machining nearly all other materials. Yet, also in comparison with unreinforced metals, there are reasons why metal matrix composites are attractive.

The composite approach is the only pathway for the production of entire classes of metallic materials. Aluminium, copper, and magnesium metals cannot be combined with significant volumes of carbide, oxide or nitride phases by conventional alloying. Unlike ferrous alloys, melts of these relatively low melting alloys have no significant solubility for the elements carbon, nitrogen and oxygen (with the exception of O in Cu). To produce a copper material containing carbon, copper and carbon must therefore be combined using processing approaches other than alloying and casting. Carbon-containing copper can be produced by infiltrating carbon (*e.g.,* fibre bundles, a porous block of graphite, or loosely packed graphite powder) with molten copper: the resulting material is by definition a composite.

The composite approach also enables significant alterations in the physical properties of metallic materials that cannot be achieved by simple alloying. Composites offer for example a means for exceeding a specific elastic modulus (Young's modulus divided by density) of 26 $J \cdot kg^{-1}$, a value exhibited by the vast majority of engineering metals, in a material that still has the basic attributes of a metal. Composites also offer the only pathway for producing materials with tailored combinations of physical properties. An example is the coefficient of thermal expansion, which can be specifically designed in the composite to match that of another, non-metallic, material such as alumina or silicon. Such a tailored coefficient of thermal expansion, coupled with a high thermal conductivity characteristic of metallic materials, is a very attractive property combination for electronic packaging applications.

Finally, metal matrix composites can offer properties that are exceptional compared with those of any unreinforced alloy. Nextel alumina fibre reinforced Al-2wt%Cu composites produced by 3M, for example, now exhibit a tensile strength on the order of 1.5 GPa and a compressive strength around 3 GPa along the fibre direction, while across the fibre direction the tensile strength remains above 200 MPa (see Table 2.8 in the next chapter). This material is only 22% more dense than aluminium and is capable of retaining its strength to higher temperatures than most aluminium alloys. This unique combination of properties makes such a composite material attractive for several specific engineering applications despite its relatively high cost.

1.4 The Development of Metal Matrix Composites

While research on metal matrix composites was pioneered in the 1960's, metal matrix composite materials have become the object of a significant and sustained R&D effort only since the early 1980's. Japan contributed significantly to their technological development with an effort that combined contributions from government, industry, and academia, reaching peak activity in the second half of the 1980's. At the same time, the "Star Wars" Strategic Defense Initiative project in the United States and later the National Aerospace Plane (NASP) project both featured relatively large and focused efforts on metal matrix composites, as did several other government-subsidized research grants, large and small, throughout the United States. Somewhat later in Europc, metal matrix composites also became the subject of significant research efforts, both at national levels and within the framework of larger European Union projects.

Metal matrix composites have caught the attention of industry to varying extents and at different times. Most aluminium producers and many automotive and aerospace companies at some point have contributed to the development of these materials. The level of contribution and effort has varied significantly from one company to another, however. Some companies have adopted a "stand-by and watch" approach to the technology. Others, like Péchiney in France, Sumitomo Chemical in Japan, and Alcoa in the United States, essentially "gave up" after contributing to the technology in a significant way. On the other hand, a number of companies, small and large, have played significant roles in the development of metal matrix composite technology for more than a decadc and remain active developers and producers of these materials; examples include Alcan, DWA, Honda, Toyota, and General Motors.

Value estimations for the technological outcome of research into metal matrix composite technology have varied significantly, depending on the estimator and also on the time of the prediction. Twenty years ago, amidst much hype about these materials, the future technological impact of metal matrix composites was believed to be very high. Aerospace was viewed as obviously hungry for new lightweight, high strength and high modulus materials such as metal matrix composites, while Toyota had eloquently demonstrated their usefulness in large-scale applications: a selectively reinforced aluminium diesel engine piston was launched in the early 1980's and has to the best of our knowledge been maintained in production (with improvements in design) to this day. For the past ten years, such estimations have on average been more sober: overly optimistic predictions that viewed metal matrix composites as destined for wide-scale usage as commonplace structural materials have disappeared. At the same time, several tangible but specialised application areas for these materials have emerged, demonstrating their usefulness.

Market studies conducted about ten years ago also varied widely: in 1990 the world-wide market of MMC was estimated at between 3 and 37 million US $, and predictions for 2000 ranged between 60 and 450 million US $ [1]. A recent study estimates the 1999 worldwide market for metal matrix composites at 103 million US $, corresponding to the production of about 2.5 million kilograms of material [2]. Forecasting is obviously difficult, particularly when dealing with a branch of technology still undergoing active development. Also, it must be noted that estimations of market size vary widely depending on the precise definition one uses to identify metal matrix composites (see Chapter 2). Irrespective of the assumptions and definitions that one uses, it is apparent from market evaluations that these materials are penetrating the marketplace, are being used in industry, and are still far from saturating their technological potential.

1.5 Statement of Purpose

We are thus currently at a stage where metal matrix composites are making a few tangible inroads into the "real" world of engineering. Yet, outside a relatively limited circle of specialists, these materials are mostly unknown. Designers do not as a rule envisage using metal matrix composites: their novelty and relative complexity add unwanted complications to an already challenging design process, information is relatively scant in many engineering materials handbooks, and there is a lack of extensive industrial experience.

These observations motivated the proposal and funding of a Thematic Network within the Brite Euram III European Union research programme. This network, titled "MMC-Assess", was constituted in Oct. 1998 and coordinated by Prof. H.P. Degischer of the Technological University of Vienna. It brought together 21 partners from industry, applied research institutions and universities, thus assembling a representative cross-section of European researchers active in metal matrix composites [3]. The Network's main goal was to increase market acceptance of metal matrix composites by collecting and evaluating information on their technological characteristics and outstanding potential for innovative applications. The present volume is an outcome of the authors' participation in this Network, which was funded in Switzerland by the Federal Office for Education and Science (BBW/OFES).

We have sought to construct a database outlining the "real world" activity in metal matrix composites. We do not aim to predict, or even to quantify, current markets for metal matrix composites. Rather, our goal is to present a landscape of the MMC industry at present: a global survey (not exhaustive, but reasonably complete) with sufficient detail to provide the reader with a useful description of the commercial activities in this new and growing branch of materials technology. To this end, we have excluded all research and development efforts and have sought to include only information about actors actively using or selling metal matrix composites on the marketplace. As for all activities of the MMC-Assess Network, we have made use of only open information, that is freely available to the general public.

1.6 Structure

Our purpose is to describe the present-day technological base of the metal matrix composites industry and disseminate public information about this technology to as wide an audience as possible. To this end, we begin by classifying and describing the main categories of metal matrix composites. We then describe the processing techniques relevant to these materials. The reader who is already cognizant of metal matrix composites can obviously skip these two next sections and move straight to the fourth, in which we present the methodology used in assembling the present database and describe its structure. We then provide, in the sixth and main section of the volume, a textual and tabular overview of producers, materials and their applications, *i.e.*, the database itself accompanied by short descriptions of the companies within it, sorted alphabetically by company. The database is included in the accompanying disk and requires Filemaker™ Pro software (version 5) to view and search using the terms outlined in Appendix B. We conclude with a brief discussion of this young industry, and provide in Appendix A a glossary of terminology commonly used in this area of technology.

For illustration, we have mostly used visual material emanating from research in our own laboratories. This was motivated by convenience, and is by no means meant to imply that our work on the subject has greater significance than that of our many colleagues.

Chapter 2

METAL MATRIX COMPOSITES

2.1 General

In their most common form, metal matrix composites are combinations of a metal with a ceramic ("cermets" was a term used about 40 years ago to designate metal/ceramic composites of high ceramic content). Other combinations nonetheless exist. Metallic reinforcements are used in several materials that are metal matrix composites by the present definition: tungsten reinforced copper, tungsten heavy alloys, and copper-niobium superconductors are examples. Metal foams can also be viewed as metal matrix composites according to this definition (the second phase is a gas).

As these examples show, there is wide latitude in dealing with the subject of metal matrix composites. We have opted for a relatively narrow and contemporary scope of study, and therefore made the decision to exclude several categories of metal matrix composites from consideration. These are:

(i) metal-metal composites. These include tungsten heavy alloys (W-Cu-Ni, W-Ni-Fe, W-Cu), Cu-Mo materials, Cu-infiltrated sintered steel powder compacts, and composite superconducting wire materials;

(ii) "carbide" or "diamond" tool materials, including WC-Co and other cermets used in the cutting tool industry, as well as impregnated diamond tool materials and other diamond tool materials ("PCD" polycrystalline diamond, electroplated diamond grinding wheels, etc);

(iii) Copper and silver matrix composites for electrical contacts and for tooling in electromachining operations;

(iv) dispersion-hardened metals, such as oxide dispersion strengthened (ODS) materials, because in such materials the reinforcement represents only a small fraction of the material by volume and therefore does not bear a significant fraction of the applied stress - rather, its role is mainly to impede dislocation motion in the matrix;

(v) metal foams, except as an application of particulate reinforced aluminium when this material is used to produce the foam (in which case the foam is simply an application of a composite surveyed herein).

These choices were made because the above composites are at a very different stage of development than the metal-ceramic composites that are the focus of the present survey. Composites in categories (i), (ii), (iii) and (iv) are well-established materials in their respective fields of application; their usage, properties, and vendors are well known and relatively stable. At the opposite extreme, metal foams are at an earlier stage of development than the metal matrix composites surveyed here. Metal foams would therefore not appear with much significance in a survey devoted solely to "real world" engineering activities, so we have excluded them from the onset. This philosophy has not been rigidly employed, however.

These relatively strong restrictions still leave a wide range of material as the subject matter of this volume. More often than not, materials of present interest consist of a light-weight metal such as aluminium or titanium reinforced by a ceramic, such as SiC particulate or SiC monofilament. The vast majority of these metal matrix composites can be sorted into four broad classes based on the form of the reinforcing phase:

(i) particulate reinforced metals (PRM);
(ii) short fibre reinforced metals (SFRM) and whisker reinforced metals (WRM);
(iii) continuous fibre reinforced metals (CFRM); and
(iv) monofilament reinforced metals (MFRM).

As these categories imply, the term "metal matrix composites" (even in the restrictive sense adopted here) designates an array of materials having a wide array of properties and attributes. Some metal matrix composites, for example, are isotropic while others exhibit strength along one direction that is about thirty times that in the perpendicular directions. Furthermore, the four categories described above are not all-encompassing: other types of composite exist. In interpenetrating phase composites ("IPC"), for example, matrix and reinforcement are both continuous throughout the material, forming two intertwined phases, alternatively concave and convex; an example is given in Figure 2.1. IPCs are significantly less common than the four basic types of composite listed above.

The diversity of metal matrix composites can be further appreciated by considering cost: commercial prices of metal matrix composites span about four orders of magnitude, from a few US dollars to several tens of thousands of dollars per kilogram. We provide in what follows an elementary description of metal matrix composites in each of these four main categories; Chapter 3 then describes how metal matrix composites are made.

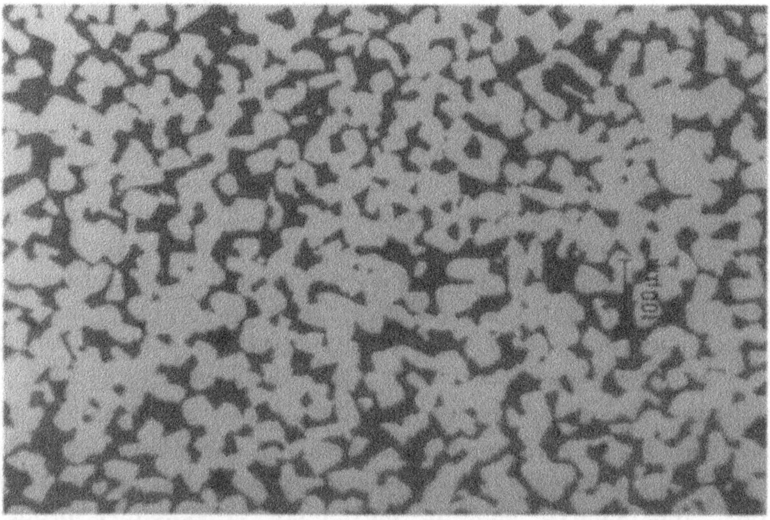

Figure 2.1 - An interpenetrating phase composite (IPC) of SiC and copper produced by infiltrating with copper a fine SiC foam produced by the process described in Ref. [1] (composite made at MIT by T. Fitzgerald).

2.2 Particulate Reinforced Metals

Particulate (or particle) reinforced metals (PRM) are combinations of a metal or alloy with a roughly equiaxed second phase that is deliberately introduced into the metal in order to improve its properties. The term "equiaxed" is a matter of definition: in simple and practical terms, it means a reinforcement that is a powder before it is combined with the matrix. To be somewhat more rigorous, we define particulate as any discrete reinforcement that does not have one dimension more than five times longer than the other two; in this way long fibers are excluded but not platelets. The composite shown in Figure 1.1 (previous chapter) is a particulate reinforced metal.

Compared with fibre-, whisker- and monofilament-reinforced composites, particulate reinforced metals offer inferior improvements in mechanical properties. Particulate reinforced metals, however, have significant advantages compared to these, which include:

(i) low cost of most equiaxed reinforcements,
(ii) simple and low-cost production processes,
(iii) isotropic composite properties, and

(iv) compatibility with most conventional metalworking processes (machining, deformation processing, welding, etc.) if the reinforcement volume fraction is less than about 40 vol. %.

These composites are thus closer to conventional metals and alloys than fibrous metal matrix composites in terms of cost, processing, and properties.

Metal matrices are most often based on aluminium; however, magnesium, copper, and steel have successfully been reinforced with particles (nickel and cobalt are typically the matrices for "hardmetal" cermet tool materials; these are as mentioned not surveyed here). Although metallic particles (chiefly tungsten) are used in some composites, particles of interest here are primarily ceramic, often made from the lighter elements: the most common examples are SiC, Al_2O_3, B_4C, TiB_2 and TiC. Graphite has also been added to aluminium as a dry lubricant and as a chip-breaking phase in machining, but has the disadvantage of being weak, resulting in composites with poor mechanical properties. Inexpensive ceramic particulate is generally produced by comminution (crushing). Powders can also be grown in solution (sol-gel processing), generally at a significantly greater cost than those that are comminuted. These premium powders are produced for the sintering industry, but they may find their way into metal matrix composite materials.

Particulate reinforced metals can be sorted into two rather distinct groups according to the volume fraction of reinforcement. High volume fraction materials are distinguished by contact between individual reinforcing elements. Reinforcement concentrations of greater than about 40 vol % are necessary to achieve a network of contacting particles; however, the transition between contacting particles and particles separated one from another by matrix depends on the particle shape and size distribution. Figure 1.1 (previous chapter) gives an example of a high volume fraction PRM with a monomodal particle size distribution, while Figure 2.2 illustrates a composite with a still higher volume fraction ceramic produced using multimodal particle size distributions.

High volume fraction PRM are typically produced by infiltrating powder preforms with molten metal. These materials appeared in the 1990's for thermal management in electronic packaging. In this application, high concentrations of reinforcement are desired, so as to match the thermal expansion of silicon or alumina while maintaining the thermal conductivity of the metal matrix. Electronic applications of particulate reinforced metals have grown steadily during the past decade.

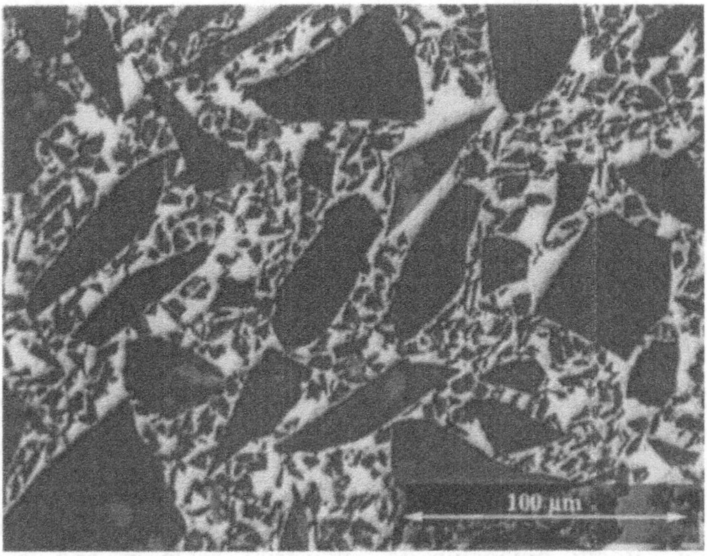

Figure 2.2 - Infiltrated SiC reinforced Al-Si matrix composite: this composite was produced by squeeze casting of the matrix into a preform of the reinforcement, itself produced using a bimodal particle size distribution to reach the high volume fraction reinforcement shown (micrograph produced at EPFL, from a composite produced at KAIST, Korea, in the laboratory of Prof. S.H. Hong).

In contrast to the high volume fraction composites, most ceramic particles are completely surrounded by matrix in low volume fraction particulate reinforced metals; Figure 2.3 provides an example of such a composite. The low fraction particulate composites are destined for structural and wear applications, and they contain less than about 30 vol. % ceramic, primarily with the goal of maintaining the ductility and toughness at levels comparable to engineering alloys.

These particulate reinforced metals have been produced for research at universities roughly since the 1960's (e.g., [2, 3]). About two decades later production of particulate reinforced aluminium followed in experimental or prototype quantities, primarily by powder metallurgy (P/M). DWA remains active after nearly three decades of producing particulate reinforced metals by P/M and several small- to medium-sized companies have entered the market more recently.

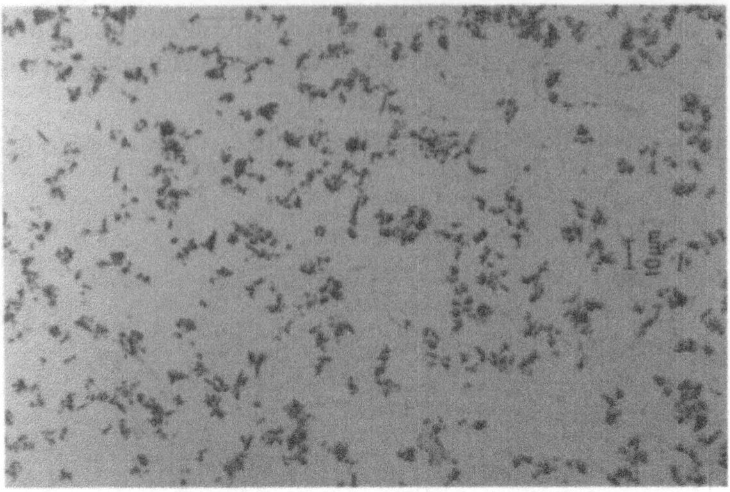

Figure 2.3 - SiC particles dispersed in magnesium (Composite produced at MIT by E. Klier using the process described in Refs. [4, 5]).

In the early 1980's M. Skibo and D. Schuster proposed a molten metal process for the production of PRM in place of the more expensive solid-state route. Being amenable to industrial scale-up and capable of producing material of good quality, the stir-casting process patented by Skibo and Schuster [6-8] produced a second impetus behind the development of PRM. Their start-up company (Dural Composites) and its process were purchased by Alcan, whose investment into this technology has established it as the largest supplier of particulate reinforced aluminium. This composite is marketed under the tradename Duralcan.

While P/M and stir-casting form the backbone of particulate reinforced metal production, additional processes have appeared. In-situ processes have been used industrially. TiB_2 reinforced aluminium produced by London and Scandinavian is one example; here the reinforcement is formed within the matrix by chemical reaction. Other processes have also been explored, and although these are not used industrially today some may assume importance in the future. Spray-casting, in which particles are sprayed into a cone of atomized molten metal drops which impinge and solidify over a growing ingot of composite, is one important example of such a process.

Particulate reinforced metals resemble unreinforced metals and alloys more than the fibre-, whisker- and monofilament-reinforced MMCs, as mentioned above. This is particularly true of low fraction PRM since even if reinforced with as much as 35 vol % ceramic they can be purchased in bulk form and processed by most conventional metalworking processes (casting, machining, deformation processing, welding, etc.). This has facilitated their entry into engineering practice.

Ceramic particulate reinforced metal matrix composites offer some ubiquitous (or nearly so) benefits compared to their matrix, generally with only modest changes in density. These benefits include:

(i) enhanced wear and erosion resistance,
(ii) increased stiffness,
(iii) higher damping, and
(iv) reduced thermal expansion.

These properties can be tailored (within a limited range of course) by adjusting the fraction ceramic in the composite. Wear resistance is dramatically enhanced with only small additions of reinforcement, while stiffness is gradually increased and thermal expansion is gradually reduced as the ceramic concentration is increased.

With regard to strength, particulate reinforced metals feature attractive compressive properties; however, their ultimate tensile strength will generally not surpass that of an unreinforced high-strength alloy of the main matrix element. For instance, present particulate reinforced aluminium composites essentially match, but do not exceed, the performance of the strongest aluminium alloys in terms of ultimate tensile strength. This becomes particularly evident when comparing the P/M composites to the rapidly solidified P/M aluminium alloys, as should in all rigor be done since the composite matrix was itself rapidly solidified.

Advantages may also depend on microstructural size: for example, if the particles have diameters smaller than about one micrometre, they impede dislocation movement and recovery in the matrix (as in dispersion hardened metals). The creep resistance of the composite material is then superior to that of the matrix alone. Other benefits over alloying can be achieved that depends more strongly on the specific material combinations. These include:

(i) increased resistance to stress-limited high-cycle fatigue,
(ii) increased hardness,
(iii) increased resistance to various high-temperature environments,
(iv) improved thermal conductivity (when reinforced with SiC or BeO),
(v) dry lubrication (when reinforced with graphite), and
(vi) lower density (when the reinforcement has a lower density than the matrix, i.e. B_4C in Al, TiC in steel).

In addition to higher cost, lower ductility and lower toughness (which also reduces their fatigue performance at high stress) are the chief drawbacks of particulate reinforced metals compared with their unreinforced matrix. These drawbacks are natural consequences of adding a brittle and very stiff second phase to a ductile metal; Figure 2.4 clearly illustrates this point. Reduced ductility and toughness have long been perceived as the chief limitations (other than cost) of ceramic reinforced metals. Progress in the design and processing of the materials, particularly over the past decade, however has brought the ductility and toughness of low volume fraction particulate-reinforced aluminium close to those of engineering aluminium alloys. These properties are therefore in general no longer considered "show stoppers" for state-of-the-art particulate reinforced aluminium in structural applications.

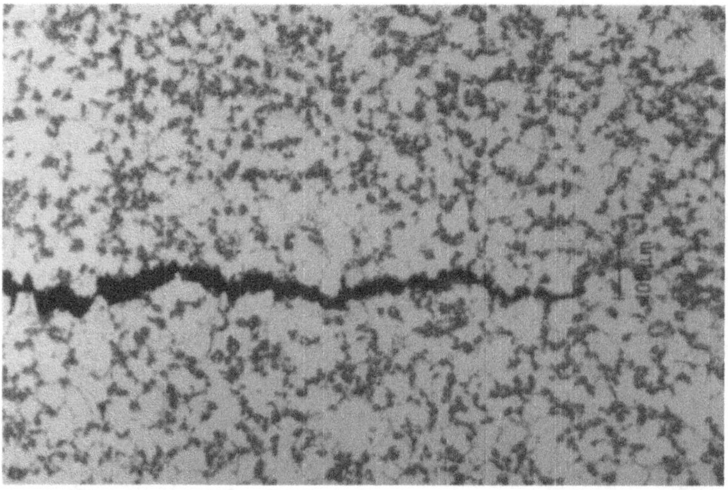

Figure 2.4 - A crack meandering its way through a sand-cast 356 aluminium alloy reinforced with SiC particles: the particles are segregated into clusters because they were pushed by solidifying matrix dendrites. This provides a preferential pathway for propagation of the crack, as shown. The tensile elongation of such a composite is, therefore, significantly lower than that of its unreinforced matrix alloy. Improved processing and microstructural design of the composite, in particular to reduce ceramic particle clustering, produces a material with improved properties.

The mechanical properties of several structural (low volume fraction ceramic) particulate reinforced aluminium composites that are currently available on the market are given in Tables 2.1 to 2.5. Tables 2.1 and 2.2 list properties of two aluminium/SiC powder-metallurgy composites, while Tables 2.3 and 2.4 lists properties of composites produced by stir-casting

Table 2.1 - Typical Properties of 6092/SiC/extruded/T6 DWA Aluminium Composites MMCs; Source: References [9, 10].

Particulate Vol.%	**17.5**					**25**				
Product Thickness (mm)	1.8	2.3	3.8	6.4	20.8	6.4	8.9	12.7	16	18.8
Density (g/cm^3)	2.80	2.80	2.80	2.80	2.80	2.82	2.82	2.82	2.82	2.82
Tensile Ultimate Strength (MPa)	510	510	490	490	462	538	545	517	524	517
Tensile Yield Strength 0.2% (MPa)	448	448	434	421	393	517	524	483	483	483
Compressive Yield Strength (MPa)	462	462	414	407	365	490	455	386	407	427
Young's Modulus, Tension (GPa)	105	105	107	108	108	123	123	122	123	121
Young's Modulus, Compression (GPa)	106	105	106	107	104	121	121	119	121	121
Tensile Elongation to Failure (%)	6	7	6	6	7	4	4	5	5	5
Shear Ultimate Strengh (MPa)	303	303	296	303	276	331	331	317	310	310
Poisson Ratio								0.29		
Fracture Toughness, K_{Ic} ($MPa.m^{1/2}$)								20.4		
Bearing Ultimate Strength e/D = 1.5 (MPa)	848	793	696	724	641	745	758	696	703	703
Bearing Ultimate Strength e/D = 2.0 (MPa)	1055	1000	924	917	800	931	965	896	903	945
Bearing Yield Strength e/D = 1.5 (MPa)	745	703	641	648	600	717	710	648	648	682
Bearing Yield Strength e/D = 2.0 (MPa)	896	827	800	786	724	855	855	779	772	834
Average CTE (ppm/°C)								15.3		
Thermal Conductivity (W/ m.°C)								164		

(Table 2.3 for extruded Al_2O_3 particulate reinforced wrought aluminium matrix composites, Tables 2.4 and 2.5 for cast SiC/Al-Si alloy composites). The mechanical properties of powder metallurgy materials are superior; however, so is their cost.

Table 2.2 - Typical Properties of 2124/SiC/25p/T4[(a)] Aerospace Metal Composites (AMC) MMCs. Source: References [9, 11].

Processing/Form	Extrusion/ Bar	Plate	Forging
Product Thickness (mm)	5 to 30	5 to 20	5 to 50
Density (g/cm^3)	2.88	2.88	2.88
Average Particle Size (μm)	3	3	3
Tensile Ultimate Strength (MPa)	690	659 [(b)]	680
Tensile Yield Strength 0.2% (MPa)	487	464	450
Tensile Elongation to Failure (%)	5	3.5	5
Young's Modulus, Tension (GPa)	115	115	115
Shear Modulus (GPa)	44	44	
Poisson Ratio	0.3	0.3	0.29
Fracture Toughness, K_{1c} ($MPa.m^{1/2}$)	19	19	
Fatigue Strength (rotating bend fatigue) R=-1, K_t=1 (MPa), 10^5 cycles	400	430	
Fatigue Strength (rotating bend fatigue) R=-1, K_t=1 (MPa), 10^7 cycles	350	300	
Brinell Hardness	210	210	
Rockwell Hardness (HRB)	95	95	
Vickers Hardness [(c)]	210	210	
Average CTE [(d)] (ppm/°C)	15.5	15.5	15.5
Thermal Conductivity [(e)] (W/ m.°C)	150	150	150
Specific Heat (J/ g.°C)	0.836	0.836	
Melting Point (°C)	min. 548	min. 548	
Solidus (°C)	548	548	
Liquidus (°C)	660	660	
Electrical Resistivity (mW.cm)	9	9	

(a) designation: AMC225xe; (b) statistical analysis gives AVG-3STD at 417 MPa; (c) pyramidal diamond 10 kg load; (d) linear 20°C (over -100 to 100°C); (e) 0 to 100°C.

Table 2.3 - Typical Properties of 6061/Al_2O_3/extruded/T6[(a)] and 2618/Al_2O_3/extruded/T6[(b)] Duralcan MMCs. Source: References [9, 12].

Matrix	**6061**			**2618**		
Particule Reinforced (vol.%)	10	15	20	10	15	20
Tensile Ultimate Strength (MPa)	352	365	372	462	469	462
Tensile Yield Strength 0.2% (MPa)	296	324	352	400	421	421
Tensile Elongation to Failure (%)	10	6	4	3.3	2.4	2
Young's Modulus (GPa)	81.4	88.9	97.2	88.3	95.1	104
Poisson Ratio				0.32		0.31
Fracture Toughness, K_{Ic} (MPa.m$^{1/2}$)	24.1	22	21.5	20.3		19.3
Fatigue Limit, 10^7 Cycles, R=-1 (MPa)	175		150			150
Density (g/cm^3)	2.81	2.86		2.88		3
Average CTE (ppm/°C) 50-100 °C	20.9	19.8		20.9	19.8	
50-300 °C	22	19.6				
50-500 °C	23	20.3				
Thermal Conductivity (W/m.°C) 22 °C	156	144		156	144	
Thermal Conductivity (W/m.°C) 150 °C	170	156				
Thermal Conductivity (W/m.°C) 200 °C	173	162				
Thermal Conductivity (W/m.°C) 260 °C	180	168				
Abrasion Resistance-Volume Loss[(c)] (mm^3)	0.19	0.15				
Block-on-ring Wear Test-Volume Loss[(d)] (mm^3)	0.02 6	0.017	0.014			
Coefficient of Friction[(e)]	0.49		0.45			
Electrical Conductivity - 22°C (%IACS)	35.2	32.7				

(a) commercial designation W6A.xxA-T6; (b) commercial designation W2F.xxA-T6; (c) ASTM G-65B sand abrasion test; (d) measurements according to ASTM G-77; (e) ASTM D-3702 thrust-washer wear test.

Table 2.4 - Typical Properties of 360/SiC/high pressure die cast[(a)] Duralcan MMCs. Source: References [9, 12].

Particulate vol.% *(b)*	**10**			**20**		
Temper	F	O	T5	F	O	T5
Tensile Ultimate Strength (MPa)	310	248	352	303	262	365
(minimum values)	*269*	*241*	*317*	*262*	*248*	*338*
Tensile Yield Strength 0.2% (MPa)	221	145	317	248	159	338
(minimum values)	*186*	*138*	*296*	*221*	*138*	*317*
Tensile Elongation to Failure (%)	0.9	2	0.5	0.5	1.5	0.3
Young's Modulus (GPa)	91	91	91	108.2	108.2	108.2
Poisson Ratio				0.293		
Rockwell Hardness (HRB)	56	40	67	73	49	73
Impact Toughness[(c)] (J)	1.4	2.7	1.4	0.7	2	0.7
Density (g/cm^3)	2.65			2.71		
Average CTE (ppm/°C) 50-100 °C	21.4			16.6		
50-300 °C	24.5			20		
50-500 °C	26.3			20.2		
Thermal Conductivity (W/ m·°C) 22°C	161			168		
260°C	159.6					
Specific Heat (J/ g·°C) 25°C	0.87			0.807		
100°C	0.915			0.898		
150°C	0.961			0.948		
Corrosion Rate - Neutral Salt Spray Test[(d)] (mm/yr)					0.3	0.3
Abrasion Resistance- Volume Loss[(e)] (mm^3)				0.19		0.21
Block-on-ring Wear Test- Volume Loss[(f)] (mm^3)				0.015		0.02
Melting Point (°C)	575-600	575-600	575-600	575-600	575-600	575-600
Electrical Conductivity - 22°C (%IACS)	32.7			24.7		

(a) commercial designation F3N.xxS-F; (b)Average particle size: 10 to 15 μm; (c) Charpy impact test, ASTM E-23; (d) after 42-day exposure; (e) ASTM G-65B sand abrasion test; (f) measurements according to ASTM G-77.

Table 2.5 - Typical Properties of 380/SiC/high pressure die cast [(a)] Duralcan MMCs. Source: References [9, 12].

Particulate vol.% *(b)*		**10**			**20**	
Temper	F	O	T5	F	O	T5
Tensile Ultimate Strength (MPa)	345	276	372	352	303	400
(Minimum Values)	*303*	*241*	*310*	*296*	*276*	*365*
Tensile Yield Strength 0.2% (MPa)	241	152	331	303	186	*(c)*
Tensile Elongation to Failure (%)	1.2	1.7	0.7	0.4	0.8	*(c)*
Young's Modulus (GPa)	93.8	93.8	93.8	113.8	113.8	113.8
Poisson Ratio	0.296			0.287		
Fatigue Limit, 10^7 cycles, R=-1 (MPa) 22 °C				130		
Same; 200 °C				60[(d)]		
Fatigue Limit, 10^7 cycles, R=0.1 (MPa) 22 °C	150			160		
Rockwell Hardness (HRB)	77	55	84	82	62	87
Impact Toughness[(e)] (J)	1.4	2.7	1.4	0.7	1.4	0.7
Density (g/cm^3)	2.76			2.82		
Average CTE (ppm/°C) 50-100 °C	19.3			16.9		
Thermal Conductivity (W/ m.°C) 22°C	124			144		
Specific Heat (J/ g.°C) 25°C	0.84			0.828		
Corrosion Rate - Neutral Salt Spray Test[(f)] (mm/yr)						0.7[(d)]
Abrasion Resistance-Volume Loss[(g)] (mm^3)				0.18		0.185
Block-on-ring Wear Test-Volume Loss[(h)] (mm^3)				0.02		0.025

(a) commercial designation F3D.xxS; (b) average particle size: 10 to 15 μm; (c) test bar fracture before yielding; (d) graphically deduced values; (e) Charpy impact test, ASTM E-23; (f) after 42-day exposure; (g) sand abrasion test: ASTM G-65B; (h) measurements according to ASTM G-77.

2.3 Short Fibre and Whisker Reinforced Metal

Short fibre reinforced metal (SFRM) and whisker reinforced metal (WRM) contain elongated reinforcements that have an aspect ratio greater than five but are not continuous. The definition of "not continuous" is somewhat of an academic issue, which can be expressed in terms of load transfer between the matrix and the fibre in the composite under stress (*e.g.*, short fibres are fibres shorter than five times the critical load transfer length defined by shear-lag analysis). In practice, however, short fibres and whiskers are easily distinguished from continuous fibres, as they are sold in bulk similar to powder or cotton, or after pre-processing into mats or preforms. By contrast, all continuous fibres currently available on the market are sold as tows of parallel and very long fibres wound on spools.

Short fibres (also called chopped fibres, or staple fibres) can be produced directly from a melt or slurry, for example by spinning in a highly turbulent atmosphere. Continuous fibres can also be cut (or "chopped") to produce short fibres. Figure 2.6 shows short fibres packed into a preform before infiltration, and held together with a silica-based binder (visible as irregularities on the surface of the fibre).

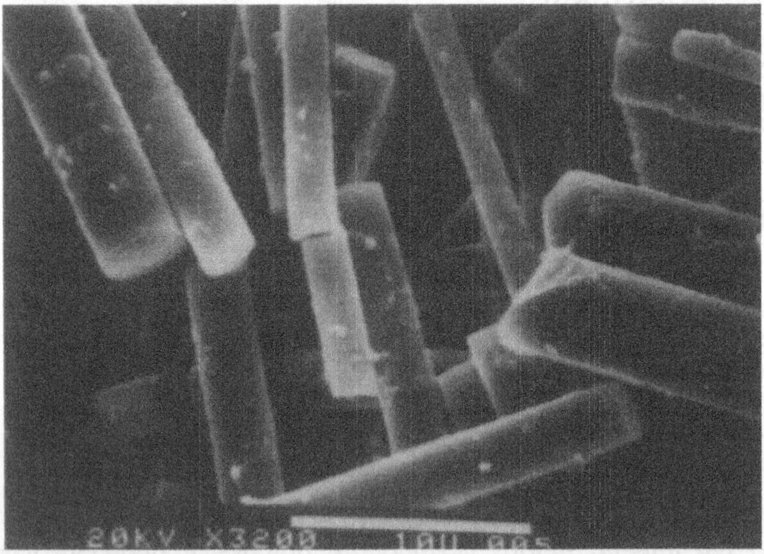

Figure 2.6 - Microstructure of an alumina short fibre preform: roughness on the fibres is due to the presence of a silica-based binder, used to bond fibres and confer cohesion to the preform, for handling prior to infiltration (Saffil™ fibres, produced by ICI, Runcorn, U.K.).

Whiskers are fibrous single crystals produced using specialized, often vapor-based, processes. One example is the VLS process, in which the crystal grows by mass transfer from vapor to solid through a drop of liquid moving with the tip of the growing single crystal [13, 14]. Another process was developed for growing SiC whiskers from rice hulls (a waste by-product of rice milling), by heating these in several steps under inert or reducing atmospheres. Whiskers have diameters on the order of 0.1 μm, whereas short fibres are at least a few micrometers in diameter. Short fibres are deliberately designed with larger diameters to reduce health hazards, while whiskers are often carcinogenic.

Anisotropy in SFRM and WRM is generally apparent due to the anisotropic orientation of the reinforcements; also, reinforcements with a greater length-to-diameter ratio result in greater anisotropy. Preferential fibre orientation is a common, and sometimes desirable, consequence of processing, for example when pre-processing of reinforcement preforms is required prior to infiltration with the metal. In many cases, these preforms and the resulting composites are so-called "random planar", *i.e.,* the fibres or whiskers are aligned randomly about one direction, but their axes lie predominantly within a single plane perpendicular to that axis. The material is then more or less isotropic in this plane, and will have five independent elastic constants. Short fibre and whisker reinforced metals can also be isotropic if truly random fibre or whisker orientations are achieved, for example by powder metallurgy or using specialized preform preparation methods prior to infiltration. This, however, is relatively rare.

Secondary processing, such as extrusion, can also produce anisotropy by aligning the fibres or whiskers predominantly along one direction. This also leads to a transversely isotropic material. Alternatively, rolling can lead to an orthotropic material. Deformation processes such as extrusion and rolling must be carefully controlled because these can cause significant breakage of the reinforcement. In fact, the damage to the reinforcements can be so great that they resemble powder after composite deformation processing. In practice, therefore, the anisotropy of SFRM and WFM is often found not to be very high, either because reinforcements are not very parallel, or because their aspect ratio is relatively small in the final composite.

Short fibre and whisker reinforced metals contain relatively low volume fractions of reinforcement (less than about 25 vol. %); typical microstructures are given in Figs. 2.7 and 2.8. These relatively low fractions and the discontinuous nature of the reinforcement imply that the reinforcement stress state is always dominated by stress transfer from the matrix. These materials then resemble particulate reinforced metals in the sense that their properties are strongly governed by those of their matrix. Indeed, SFRM and WRM are often grouped with PRM into a single class of metal matrix composites called "discontinuously reinforced metal matrix composites"; the abbreviation DRA for "discontinuously reinforced aluminium" is common.

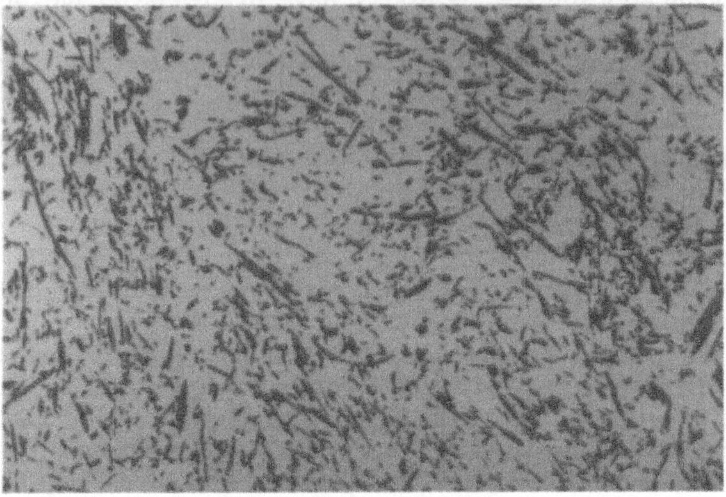

Figure 2.7 - Saffil™ alumina short fibre reinforced aluminium composite (produced by electromagnetic infiltration at MIT by R. Andrews).

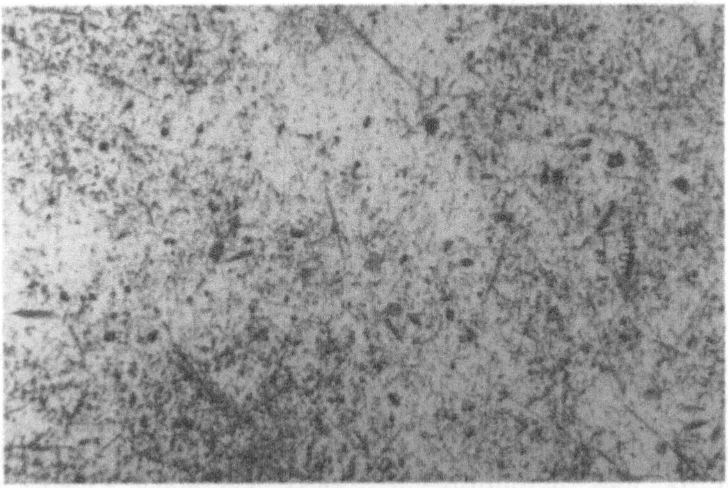

Figure 2.8 - Silicon nitride whisker reinforced Al-4.5 wt. pct. composite (produced by pressure infiltration at MIT by P. Méténier).

Property characteristics of short fibre and whisker reinforced metals, as well as their advantages and drawbacks over metal and alloys, are therefore broadly similar to those of metals reinforced with particles. Mechanical properties are not systematically improved over particulate reinforced composites, for example due to damage incurred by the reinforcement during secondary processing. Still, some improvement in mechanical properties does result from the greater capacity of elongated reinforcements to carry load transferred from the matrix compared with equiaxed (except platelet) reinforcements. In addition, being single-crystalline, whiskers are generally inherently stronger than low-cost (comminuted) particles. SFRM and WRM can therefore display somewhat superior properties than an equivalement PRM composite at a fixed volume fraction of reinforcement, particularly if the fibres or whiskers are aligned along the highest principal stress axis.

Because infiltration is often used to produce these composites and because short fibres and whiskers are readily prepared as cohesive preforms of controlled volume fraction, they can easily be placed within larger components for selective reinforcement, *i.e.*, for reinforcing certain areas only. Practical examples include the area surrounding the upper ring groove or the crown of diesel engine pistons, and cylinder liners of automotive engine blocks.

2.4 Continuous Fibre Reinforced Metals

Continuous fibre reinforced metals (CFRM) contain fibrous reinforcements that are available as tows of several hundreds or thousands of fibres. These fibres have diameters on the order of 10 μm (in practice between 5 and 20 μm), and they are produced simultaneously before being wrapped as a tow onto a bobbin.

Figures 2.9 and 2.10 give examples of the microstructure of continuous fibre reinforced metals. Since continuous fibres can be exceptionally strong and stiff, appropriately designed and processed metals reinforced with continuous fibres feature high strength and stiffness in the fibre directions and are highly anisotropic.

Continuous fibres used to reinforce metals include pure Al_2O_3 fibres from 3M or Mitsui Mining, "Nicalon" and "Tyranno" pre-ceramic polymer-derived SiC fibres from Nippon Carbon and Ube Industries respectively, 85% alumina-15%silica fibres from Sumitomo Chemical, and carbon fibres of various types (high-modulus, high-strength, PAN-derived, pitch-derived) from several manufacturers worldwide. Silica-containing ceramic fibres, such as the polymer-derived SiC and alumino-silicate fibres, have a lower Young's modulus than pure Al_2O_3 fibres.

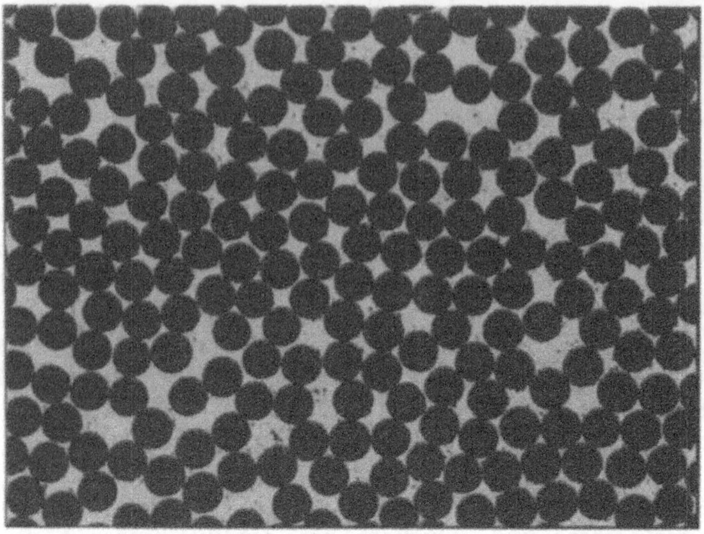

Figure 2.9 - Nextel 610™ alumina fibre reinforced pure aluminium. The scale is given by the fibre diameter, which is 12 μm (composite produced by infiltration at EPFL by B. Moser).

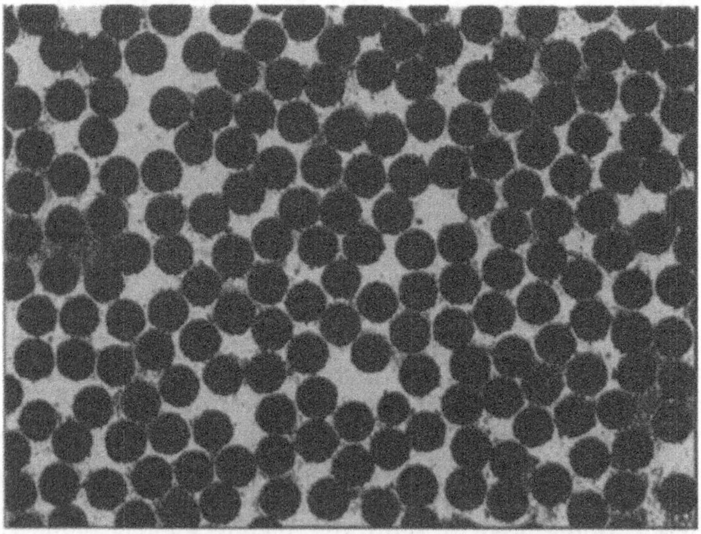

Figure 2.10 - Same composite as in Fig. 2.9 (Nextel 610™ alumina fibre reinforced pure aluminium), with the difference that small ceramic particles were used as spacers between the fibres during processing. In the final composite, the ceramic particles are embedded within the matrix, and the fibre volume fraction is lower than in Fig. 2.9. The scale is given by the fibre diameter, which is 12 μm (composite produced by infiltration at EPFL by B. Moser).

Table 2.6 gives principal properties (modulus, strength and density) of some metal-compatible fibres available on the market: many more exist, each manufacturer often offering a range of several fibres of a given type. As an illustration, Table 2.7 provides a more detailed listing of properties for the range of fibres produced by 3M. Fibre prices vary significantly; typical prices are of the order of 100 US $ per kg. More data and information on fibres for reinforcement of metal matrix composites can be found in Refs. [15-19]. Metallic matrices reinforced with continuous fibres are generally aluminium alloys, although copper and magnesium can also be used.

Table 2.6 - Some fibres that can be used for the reinforcement of metals. Data from Refs. [15-18, 20] and Table 2.7. Note the wide range of variation for carbon fibre properties (fibres with properties intermediate to those given here also exist):of course such property variations are also accompanied by significant price differences.

Fibre	Longitudinal Young's modulus (GPa)	Average tensile strength (GPa)	Density ($g \cdot cm^{-3}$)
Toray T300 PAN Carbon Fibre	230	3.53	1.76
Toray T1000G PAN Carbon Fibre	294	6.37	1.8
BP-Amoco P-25 Pitch Carbon Fibre	160	1.4	1.9
BP-Amoco K-1100 Pitch Carbon Fibre	930	3.1	2.2
Nippon Carbon Nicalon NL-200 SiC Fibre (56.6%Si, 31.7 %C, 11.7%O)	220	3.0	2.5
Ube Industries Tyranno Lox-M SiC Fibre (55.4%Si, 32.4 %C, 10.2%O)	187	3.3	2.5
Mitsui Mining Almax (99.9% Al_2O_3)	320	1.8	3.6
3M Nextel 610 (99% Al_2O_3)	373	3.2	3.8
Sumitomo Chemical Sumica (85%Al_2O_3 - 15%SiO_2)	250	2.6	3.2

Table 2.7 - Properties of 3M Nextel™ Fibres. Source: References [21, 22].

Fibre designation	Nextel 610	Nextel 650	Nextel 720	Nextel 312	Nextel 440	Nextel 550
Chemical Composition Al_2O_3 (wt%)	> 99%	89	85	62.5	70	73
Y_2O_3 (wt%)		1				
ZrO_2 (wt%)		10				
SiO_2 (wt%)			15	24.5	28	27
B_2O_3 (wt%)				13	2	
Crystal Phases	α-Al_2O_3	α-Al_2O_3 +cubic ZrO_2	α-Al_2O_3 +mullite	mullite+ amorph.	γ-Al_2O_3 +mullite+ SiO_2 (amorph.)	γ-Al_2O_3 + SiO_2 (amorph.)
Density (g/cm^3)	3.9	4.1	3.4	2.7	3.05	3.03
Melting Temperature (°C)	2000	1710	1800	1800	1800	1800
Diameter (μm)	11 .5	10 to 12	12 .5	10 to 12	10 to 12	10 to 12
Tensile Strength[(a)] (MPa)	3100	2 550	2 100	1 700	2 000	2 000
Longitudinal Young's Modulus (GPa)	373	358	260	150	190	193
Weibull Modulus	11	7	7			
Roving Strength Max. Temp. Limit [(b)] (°C)	1 300	1 400	> 1 400	950	1 200	1 300

(a) 25.4 mm gauge length, (b) 70 % strength retention; strain rate = 8 10^{-3} s^{-1}.

The small diameter of continuous fibres allows them to be easily bent to relatively small radii of curvature. The maximum elastic elongation experienced at the surface of a fibre that is bent to radius R is:

$$\varepsilon_{max} = r/R \qquad (2.1)$$

where r is the radius of the fibre. Hence, if the strain to failure of the fibre material (for example Al_2O_3) is on the order of 1%, a 10μm fibre can be bent to a radius as small as 1mm without breaking the fibre. Textile weaving and braiding techniques can then be applied to pre-process preforms of various fibre arrangements. Woven sheets containing fibre tows aligned along two perpendicular directions and stacked within the composite are typical arrangements in polymer matrix composites. Three-dimensional weaving can also be employed to produce complicated profiles. Alternatively, fibre tows can be wound to produce two- or three-dimensional shapes such as hoop-reinforced rings.

Fibre weaving, which is often used in making polymer matrix composites, is seldom used in making metal matrix composites. One reason is that metals are far stronger than polymers, such that (i) sufficient composite strength advantages over unreinforced matrices are only observed with all fibres aligned in one direction, and (ii) transverse strength is less of an issue than with fibre reinforced polymers. The second reason is that cross-woven fibres tend to cause premature failure of fibre reinforced metals.

Continuous fibre composites are generally produced by liquid metal infiltration of the fibre tows or preforms. Ceramic particles may be incorporated into the fibre tows as a means of evenly spacing the fibres, a process called "hybridisation" of the preform. Figure 2.10 shows the microstructure of a composite produced using this technique, which can be contrasted with the composite produced from a non-hybridised preform shown in Figure 2.9, in which fibres touch one another and are packed to a higher volume fraction. Secondary deformation processes are as a rule of thumb not compatible with CFRM because any substantial deformation of the composite damages the fibres.

The strong anisotropy displayed by continuous fibre reinforced metals calls for specialized approaches towards designing reinforced components. In particular, proper distribution and local orientation of the fibres within the component, and proper design of the component shape, are necessary to achieve high performance. This fact is sometimes overlooked in the development of composite prototypes since both processing and design require significant skill and know-how. As a consequence, the performance of these materials can be significantly under-utilised and misjudged in prototypes and trial applications, particularly when components are not simple shapes such as beams, tubes, rods, or wires.

The fibres in continuous fibre reinforced metals must elongate to the same tensile strain along their axis as the component they reinforce; hence, the role of the matrix in determining the average load borne by the reinforcement is (to first order) secondary. In addition, in order to achieve significant strengthening, reinforcement volume fractions are high in continuous fibre reinforced metals (about 40 vol. % or greater). Thus, the mechanical properties of the composite along the fibre direction are strongly fibre-dominated. The composites can therefore display very high strength and stiffness in this direction. Across the fibres, on the other hand, the matrix plays a dominant role since the load borne by the fibres is largely transmitted through the matrix; therefore, the properties are similar to those of discontinuous composites in transverse directions. Such pronounced anisotropy distinguishes continuous fibre reinforced metals from their particulate, short-fibre or whisker reinforced counterparts, and of course also from unreinforced metals and alloys.

Composites of high-modulus metal-compatible fibres such as alumina or carbon in an aluminium-based matrix will generally have a longitudinal Young's modulus exceeding 200 and 300 GPa respectively, compared to 70 GPa for aluminium and its alloys at room temperature (Table 2.6). The density of such a composite with 50 vol % fibres, however, will increase by less than 25% compared to aluminium, about 3.3 g cm^{-3} compared to 2.7 g cm^{-3}. These fibres have a tensile strength that is often greater than 3 GPa, resulting in composite tensile strengths greater than 1.5 GPa with 50 vol % fibres or more if the fibre tows retain their properties in the composite. Composites with significantly lower concentrations of fibres have strengths that are not significantly greater than high-strength aluminium alloys, although the stiffness of the composite will still be substantially greater. Table 2.8 gives properties of 3M's Nextel 610™ reinforced aluminium matrix composites to illustrate these points.

The mechanisms of tensile fracture along the fibre direction in continuous fibre reinforced metals are presently not fully understood. From work in the late 1970's and early 1980's (predominantly from Japan), however, it is known that the strength of the composite is strongly reduced if:

(i) the matrix contains brittle second phases (such as Si or Al_2Cu in aluminium) in contact with the fibre, or
(ii) there is significant chemical reaction between matrix and fibre (as is generally found between carbon fibres and aluminium, which react to form the compound Al_4C_3).

Both effects create effective notches or precracks along the fibre surface that are larger than the defects initially present within the fibre. Since brittle reinforcing fibres are strong as a result of being defect-free, such effective notches or precracks significantly reduce their tensile strength, by factors of two or more. The spatial distribution of parallel fibres also influences the strength of continuous fibre reinforced metals:

fibres that touch one another provide paths for propagation of a crack from one fibre to the next. Thus, small ceramic particle spacers used to separate fibres within metal matrices (hydridisation, Figure 2.10) tend to increase the composite strength at given volume fraction fibres.

Table 2.8 - Typical Properties of 3M- Nextel™ 610 α-Al_2O_3 Fiber Reinforced Aluminium and Al-2%Cu Composites with Parallel Continuous Fibres. Source: References [23-25].

Matrix		Pure Al	Pure Al	Al - 2wt% Cu (T6 temper)
Reinforcement Vol.%		**40**	**60**	**60**
Tensile Strength (MPa)	L	1240	1600	1500
Compressive Strength (MPa)	L		1700	3400
Transverse Flow Stress at 1% Strain (MPa)			120	280
Shear Strength at 2% Strain (MPa)	(LT)		70	140
Young's Modulus (GPa)	L	165	240	240
Young's Modulus (GPa)	T		130	160
Shear Modulus[(a)] (GPa)			48-57	45-46
Poisson Ratio	L		0.35	0.35
Poisson Ratio	T		0.19	0.23
Fatigue Strength[(b)], 10^7 Cycles, R= -1 (MPa)	L		400	580
Fatigue Strength[(b)], 10^7 Cycles, R= -1 (MPa)	T		40	70
Fatigue Strength[(b)], 10^7 Cycles, R=0.1(MPa)	L		800	700
Fatigue Strength[(b)], 10^7 Cycles, R=0.1(MPa)	T		40	70
Density (g/cm^3)		3.22	3.4	3.4
CTE[(c)] (ppm/°C, 25-500 °C)	L	6.6	7	9 (estimated)
CTE[(c)] (ppm/°C, 25-500 °C)	T		16	16 (estimated)
Electrical Resistivity (mW.cm)		5.3		
Electrical Conductivity (%IACS)		34		

L : longitudinal, *i.e.,* measured along the fiber direction. T: transverse, i.e., measured perpendicularly to the fiber direction; (a) measured parallel and across fibres; (b) graphically deduced values; (c) average 25-500°C.

An important parameter in fibre reinforced metals is the strength of the interface between the matrix and the fibres. It is well known from the theory of composite materials that a weak fibre/matrix interface will prevent the propagation of a crack in a plane perpendicular to aligned fibres. Weak interfaces prevent failure of one fibre from being transmitted to its immediate neighbors, effectively "diluting" among many fibres, by fibre-matrix delamination, the stress concentration that is created around one broken fibre in the composite. Current scientific literature thus distinguishes between the two extremes of "local load sharing", in which one broken fibre transmits all the load it carries to its immediate neighbors only (an unfavorable situation since these are then very likely to break, causing in turn their neighbors to break and so on until total failure of the composite), and "global load sharing", where an absence of stress concentration leads to equal sharing among all other intact fibres in the composite of the load shed by one broken fibre [26, 27]. In short, by dilution of stress concentrations, weak fibre/matrix interfaces tend to produce parallel continuous fibre composites that are strong and tough along the fibre direction.

This strategy is used to produce fibre reinforced composites which, despite having a brittle matrix such as epoxy or glass, are strong and tough along the fibre direction - it is also used by Nature in wood. With continuous fibre reinforced metals, it has repeatedly been demonstrated to be successful. For example, a fibre coating of turbostratic graphite, which is very weak along the fibre radius direction (i.e., across the fibre surface), will result in a strong composite along the fibre direction, essentially regardless of the matrix composition and microstructure (*e.g.*, [28]). A similar strategy is used in current SiC monofilament reinforced metals (see Section 2.5).

When a ductile and tough metal is the matrix of a fibre-reinforced composite, this strategy is however not always necessary: strong and tough fibre reinforced metals can also be produced with strong matrix/fibre interfaces (*e.g.*, the composites in Table 2.8). This has the important implication that unidirectional fibre reinforced metals can be very attractive, not only because they are very strong along the fibre direction, but also because they are simultaneously about as strong as a conventional metal alloy in the transverse direction (a few hundred MPa in tension with a matrix of aluminium). This can provide interesting opportunities in many design situations, *e.g.*, in rapidly rotating components [29].

That impressive properties can be obtained in fibre reinforced aluminium without the use of weak interfaces is shown by 3M's latest generation of fibre reinforced aluminium [30]. With strong interfaces, these materials show longitudinal tensile strengths above 1.5 GPa combined with transverse tensile strengths near 300 MPa, Table 2.8. In contrast, using the weak interface approach, the transverse tensile strength in similar composites with about the same longitudinal tensile strength drops to about 40 MPa [31]. The matrix composition and microstructure are important in this regard (Table 2.8), as is the fibre distribution in the composite; however, although guidelines exist (see above) there is not yet a fully established methodology for composite strength optimization based on microstructural control of the matrix. Some research groups and composite producers have therefore used systematic experiments to specify aluminium matrix chemistries tailored to yield optimized composite properties for a given composite production process and fibre type. Results can be quite interesting and show significant improvements in composite strength (*e.g.*, [20, 32-35]).

Special mention must be made of carbon fibres and their advantages and drawbacks compared to other metal-compatible fibres. These advantages include:

(i) comparatively low cost,
(ii) high strength and/or stiffness (Table 2.6),
(iii) low density (Table 2.6), and
(iv) a negative coefficient of thermal expansion along the fibre direction.

This last characteristic implies that zero thermal expansion materials can be produced by combination of fibre and metal with mutually compensating thermal expansions. This unique feature can be put to advantage in applications that require extreme dimensional stability, for example in space applications. Examples include beam structures and optical gimbals, that must not deform despite significant temperature fluctuations and gradients typical of space applications. Some pitch-derived carbon fibres ("P-100") also display another unique property: the thermal conductivity along the fibre axis is on the order of 1'000 $Wm^{-1}K^{-1}$, compared with 400 $Wm^{-1}K^{-1}$ for pure copper; however, these fibres cost around 2'000 US $/kg.

Carbon-fibre composites can be produced with matrices of aluminium, magnesium, and copper. Aluminium reacts with carbon to form the carbide Al_4C_3, which results in poor composite properties when it appears along the fibre/matrix interface; however, this reaction can be sufficiently controlled during processing to produce material with attractive longitudinal tensile strength and modulus. Carbon fibres, however, are inherently weak perpendicular to the fibre axis. Properties of carbon fibre reinforced composites along essentially all directions other than the fibre axis, and also in compression along the fibres, are therefore not very attractive compared to achievements with a strongly bonded ceramic fibre (in terms of both modulus and strength). Carbon fibres also conduct electricity; as a consequence, there is galvanic coupling between fibre and matrix. Corrosion of carbon fibre reinforced aluminium or magnesium in wet or moist environments is therefore often very severe. For the reinforcement of aluminium, carbon fibres thus have both advantages and drawbacks compared with ceramic fibres.

2.5 Monofilament Reinforced Metals

Monofilament reinforced metals (MFRM) are essentially the same as CFRM, except that monofilaments have a considerably larger diameter, on the order of 100 μm. Also, while continuous fibres are available in tows, monofilaments are available as individual elements wound on a spool. This is in part because of their size, but also because they are typically produced by vapor deposition techniques.

Three principal monofilaments have been used to produce metal matrix composites:

(i) boron monofilaments,
(ii) silicon carbide monofilaments, and
(iii) sapphire (nominally) single crystal monofilaments.

Table 2.9 - Typical Principal Properties of Monofilaments that can be used for the Reinforcement of Metals (note: these data can vary depending on production time and source), from Refs. [15, 36, 37].

Monofilament	Longitudinal Young's modulus (GPa)	Average tensile strength (GPa)	Density g/cm^3
Boron Monofilaments	400	3.8	2.54
ARC/3M Trimarc I SiC Monofilament	427	3.5	3.3
ARC/3M Trimarc II SiC Monofilament	400	4.2	3.0
DERA-Sigma 1140+ SiC Monofilament	420	3.2	3.4
Textron SCS-6 SiC Monofilament	395	4.2	3.0
Textron Ultra-SCS SiC Monofilament	370	5.6	3.0

Table 2.9 gives principal properties for the former two monofilament types.

Boron monofilaments, typically 140 μm in diameter, are an attractive reinforcement for aluminium, producing tough and strong composites with good transverse strength. Cargo bay struts of the NASA Space Shuttle are produced from boron monofilament reinforced aluminium and are still in service, representing one of the earliest successful applications of MMC technology. Boron monofilaments are produced by chemical vapour deposition from a gaseous mixture of boron trichloride and hydrogen onto a smaller "core" fibre of tungsten or carbon. For improved chemical compatibility with the matrix, boron monofilaments can be coated with silicon carbide (called "Borsic" monofilaments) or boron carbide.

Silicon carbide monofilaments are also produced by chemical vapor deposition onto a carbon or tungsten core (typically from CH_3SiCl_3 gas). The surface of SiC monofilaments is then coated with at least one layer of turbostratic graphite a few micrometres thick. This layer is essential in producing the very high tensile strengths of these monofilaments (as high as 5.6 ± 0.6 GPa for Textron's "Ultra SCS") as it blunts or deflects surface cracks.

The monofilament is also protected by the graphitic layer during handling, but this layer has the consequence that the matrix and monofilament will debond at relatively low stress. Therefore, transverse properties of the composite are lower than those of the matrix alone. Additional layers may also be present at the surface of SiC monofilaments; one example is a coating that increases in silicon concentration from the graphite layer to compositions near those of SiC or Si at the external surface. This layering, which may be repeated, acts to protect the reactive carbon layer from attack by chemical interaction with the matrix during composite consolidation.

Silicon carbide monofilaments are typically 100 to 140 μm in diameter, and comprise a carbon or tungsten core about 15 to 30 μm in diameter, a principal region of columnar SiC grains (at times in two layers of somewhat different composition because the wire is produced in two successive reaction chambers), and an outer coating of at least one layer of turbostratic graphite about 4 μm thick. Figure 2.11 shows a cross-section through a monofilament, displaying its inner structure. SiC monofilaments are presently the only viable reinforcement for titanium or titanium aluminide.

Monofilaments of essentially continuous Sapphire single crystals with a diameter around 150 μm were formerly available commercially under the trade-name Saphikon®. Although of lower strength than B and SiC monofilaments (about 2.6 GPa), the sapphire monofilaments had the advantage of chemical inertness in some matrices, and a higher thermal expansion coefficient which lowered thermal mismatch stresses between matrix and fibre.

Figure 2.11 - Cross-section through SCS-2 SiC monofilaments produced by Textron (then AVCO) in the late 1980's, embedded in a matrix of Al-4.5%Cu: the carbon core, the transition layer to the two main SiC regions, these regions and the thin outer coating can all be distinguished by etching to reveal differences in carbon concentration. The scale is given by the monofilament diameter, which is 140 *μ*m (from work reported in Ref. [38]).

All CVD monofilaments are expensive since they are solids produced from a gas that is not inexpensive. Boron and SiC monofilaments cost about 2'000 and 10'000 US $ per kg respectively (for comparison, continuous ceramic fibres and less-expensive varieties of PAN-carbon fibres cost respectively a few hundred and a few tens of dollars per kilogram) [23]. This is the main limitation to widespread use of monofilaments and monofilament reinforced metals.

The large diameters of monofilaments lead to limited bending flexibility: with a diameter of 150 μm, a surface strain of 1% is reached when the monofilament is bent to a radius of curvature of about 8 millimeters, Eq. (2.1). This places tangible limits on the composite shapes that can be produced. On the other hand, a large reinforcement has distinct advantages:

(i) each monofilament can be handled individually, laid-up, and positioned with precision during the composite production process; thus the geometrical distribution of monofilaments within the final composite can be controlled better than with continuous fibres;
(ii) there is less interface area per unit volume of composite, such that some degree of matrix-fibre interaction is tolerable;
(iii) monofilament reinforced metals typically have attractive longitudinal mechanical properties since the toughness of the material in longitudinal failure increases with fibre radius regardless of failure mechanism [39]. Boron-aluminium, although expensive, is a mechanically attractive material largely for this reason.

For relatively non-reactive metals, including Al, Mg, and Cu, the high price of monofilaments compared with continuous fibres has meant that R&D has mostly focused on the latter. Titanium and titanium aluminides, on the other hand, are too reactive to be combined with continuous fibre reinforced metals using processes such as infiltration: continuous fibres are chemically attacked and degraded by titanium. Monofilaments are, therefore, the reinforcement of choice for the production of high-performance titanium-containing matrices.

The potential benefits of reinforcing titanium with monofilaments are apparent if one considers the remarkably high intrinsic strength and modulus of commercially available SiC monofilaments. Titanium-based matrices, including intermetallics, reinforced with SiC monofilaments have received considerable attention through several large research and development programs (NASP, aimed at building a transatmospheric aircraft, TMCTECC and IHPTET aimed at the production of improved turbine engines in the United States, Brite-Euram in Europe). Ti-SiC composites are also among the best-qualified of all metal matrix composites, a fact that is apparent in the volume of the U.S. Military Handbook devoted to metal matrix composites (MIL-HDBK-17 Vol. 4): nearly all data presently in this handbook are for Ti-SiC composites.

In addition to their attractive strength and stiffness, SiC monofilament reinforced titanium and titanium-aluminide alloys have the important attribute that they remain strong at temperatures up to 400 - 600°C (depending on the system). Some of these properties and the employed matrices are presented in Table 2.10. Their exceptional properties and their low density (near half that of competing alloys such as IN100) make these composites attractive for applications in aerospace, in particular for the fabrication of medium-temperature-stage components of turbine engines for propulsion. Target or current applications include integrated blade and ring rotors ("blings"), and a divergent nozzle actuator rod produced by Atlantic Research Corp. (ARC) for the F119 engine of the USAF F-22 fighter aircraft.

Table 2.10 - Principal Properties of Monofilament Reinforced Metals, from Ref. [37] and MIL-HDBK-17 Vol. 4.

Matrix/ Monofilament	Longitudinal Young's modulus (GPa)	Average tensile strength (GPa)	Transverse Strength (MPa)
33 vol. pct. SCS6/ Ti-6Al-4V	-	1.5 (20°C) 1.3 (450°C)	-
33 vol. pct. SCS6/ Ti-24Al-11Nb	-	1.3 (20°C) 1.1 (600°C)	200 (20°C) 150 (600°C)
33 vol. pct. SCS6/ Ti-22Al-23Nb	-	1.6 (20°C) 1.5 (600°C)	400 (20°C) 300 (600°C)
35 vol. pct. SCS6/ Ti-15V-3Cr-3Al-3Sn	190	1.4 (20°C) 1.5 (425°C)	-
35 vol. pct. Ultra-SCS/ Ti-22Al-23Nb	-	2.0 (20°C) 1.7 (600°C)	-

Chapter 3

PROCESSING METAL MATRIX COMPOSITES

3.1 General

By the definition adopted here (Section 1.1), all metal matrix composites are produced by artificially combining reinforcements with a metallic phase, "artificially" meaning that the two phases preexist the material before being combined. "Making a composite" is therefore the act of combining these two preexisting phases, matrix and reinforcement, into a new material, the composite (as illustrated in Figure 1.1). This we call "primary processing" of the composite. It is distinct from "secondary processing" where the composite is deformed, shaped, hardened, machined, coated, or joined (with itself or with another material) after it is created. Preliminary processing steps may also exist upstream of primary processing; these we call "pre-processing". This of course includes making the reinforcement and the matrix, but may also comprise steps such as arranging these in various configurations. Examples include packing short fibres held together with a silica binder into a preform for subsequent infiltration, or creating powdered matrix cloth using an organic binder before solid-state consolidation with monofilaments.

An overview of manufacturing processes for metal matrix composites is provided in Figure 3.1 (*in-situ* processes are excluded since they are so numerous that a separate chart could be built for these only). In this diagram, the processes are separated into (vertical) classes by the physical state of the metal matrix during primary processing: solid-state and liquid-state processes thus form the two main categories. Remaining processes form a single category that we define as "deposition processes", where the matrix is coated on individual reinforcement elements before consolidation of these "elementary composites" using solid state bonding processes. The pre-processing, primary and secondary processes are sorted along the horizontal direction from top to bottom. The variety and breadth of processing methods is evident.

3.2 Primary processing of metal matrix composites

In all except *in-situ* processes, producing a composite consists in bonding together, at the atomic level, a premade reinforcement with a continuous metal or alloy. This task is far more involved than simply "gluing" two phases together: the fine scale of the reinforcement (usually

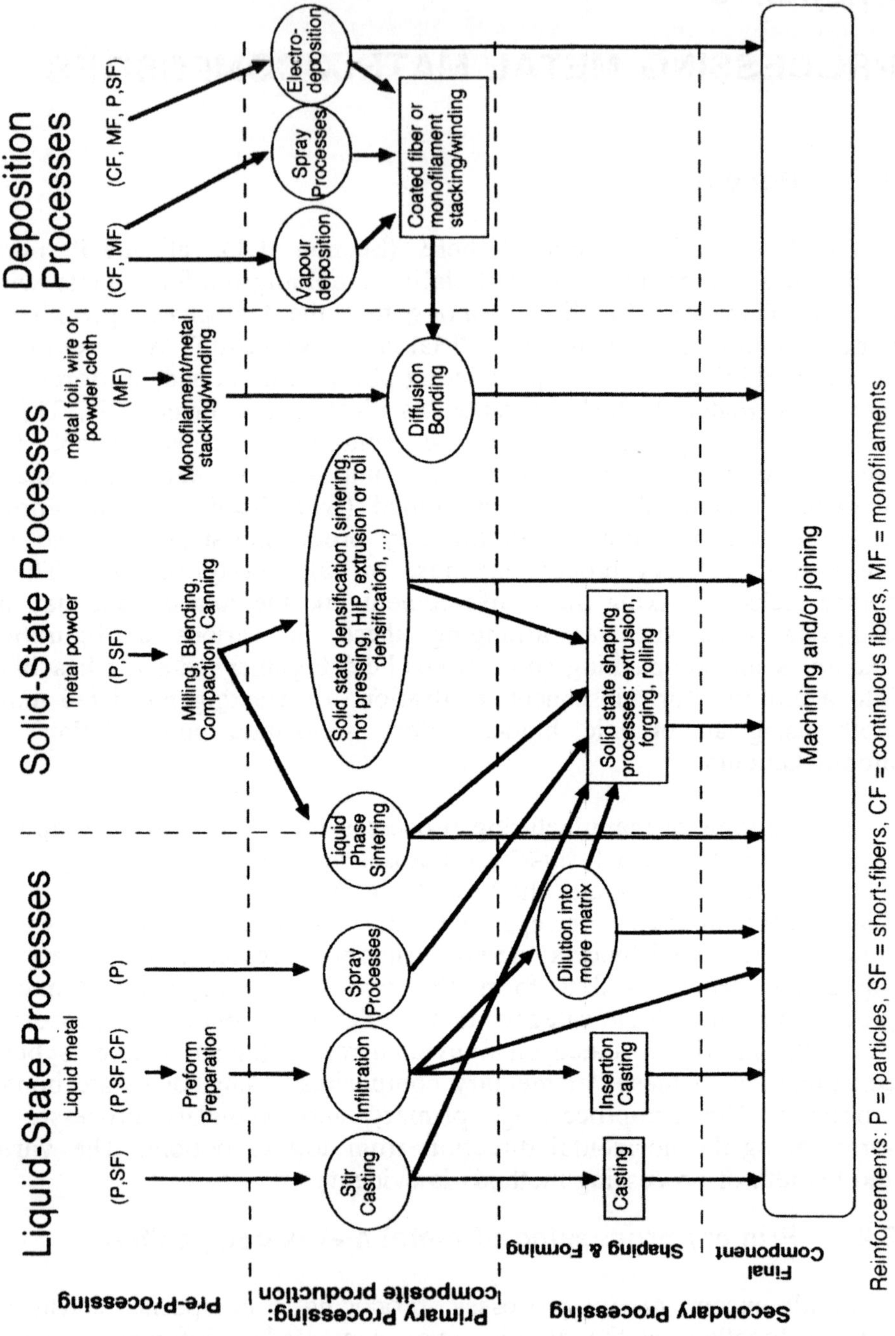

Figure 3.1 - An overview of metal matrix composite processing methods.

between 0.5 and 200 μm in diameter), coupled with the desire to combine the two materials free of defects such as pores or unwanted reaction phases, raises various critical issues. These issues depend on the class of process that is used to produce the composite.

a - Liquid metal processes

Primary processing of continuous fibre reinforced metal must insert metal into every pore of a bundle of thousands of fibres roughly 1 to 100 μm in diameter without leaving porosity and without breaking the (generally brittle) fibres. The obvious solution, for this and several other composites, is to flow low-viscosity liquid metal into and through the narrow space between individual reinforcement fibres, a process that can also be used with whiskers, particles and monofilaments [1,2].

Three of the most important metal matrix composite production processes use liquid metal:

(i) infiltration consists of preparing a porous "preform" of the reinforcement and infiltrating its pores with the molten metal as shown schematically in Figure 3.2 [3],
(ii) stir-casting consists of stirring reinforcement particles or short fibres into molten metal as shown in Figure 3.3 [4], and
(iii) spray casting consists of disintegrating a stream of liquid metal into droplets and injecting reinforcement particles that are entrapped by the molten droplets so that both deposit together to form a composite [5], Figure 3.4.

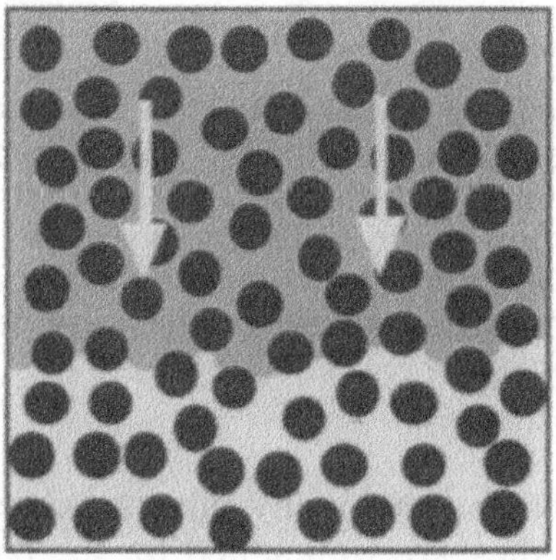

Figure 3.2 - A schematic description of the infiltration process: the molten matrix is injected into a porous "preform" of reinforcing phase elements, yielding a composite after infiltration of all open pore space and matrix solidification.

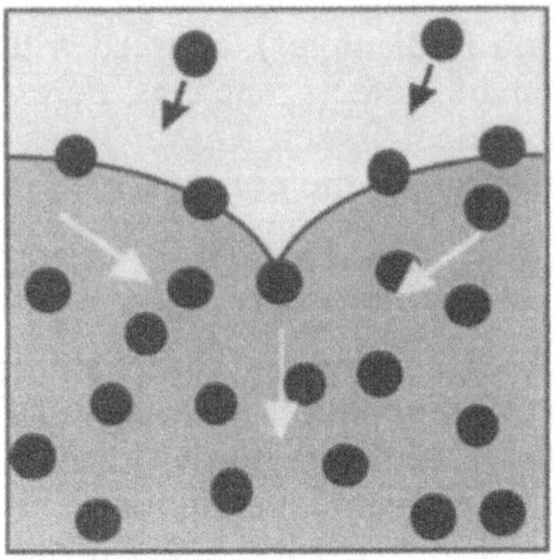

Figure 3.3 - A schematic description of the stir-casting process: particles are incorporated into a stirred bath of molten matrix; later, the composite is cast and solidified.

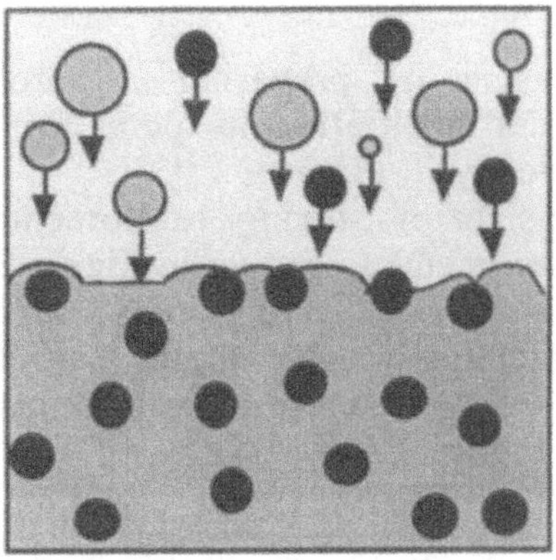

Figure 3.4 - Schematic illustration of the spray-casting process: solid reinforcement particles (black) and molten matrix droplets (grey) are sprayed together onto a chill. As the composite builds up, it serves itself as the chill over which more composite is built by impingement and solidification of the matrix, into which the reinforcement is incorporated. Some aspects of the process vary, for example the manner in which the particles are sprayed into the cone of matrix droplets.

Liquid metal processes have many advantages: they are rapid, molten metal is inexpensive, they are well suited for the production of parts with complex shapes, and they are often easy to scale up. There are of course also difficulties encountered when combining liquid metal and ceramic reinforcements.

Capillary forces create one difficulty in combining liquid metals with reinforcements, because these rarely favour the process [6-8]. From an energetic standpoint metal generally does not "want" to bond to non-metals. Like water on some surfaces (*e.g.*, a freshly waxed car hood), molten metal beads up instead of coating or "wetting" the surface of most reinforcements, Figure 3.5. Water-repellent textiles provide a simple analogy, as these are made with fibres that are poorly wetted by water: ceramic reinforcements are generally "metal-repellent". One therefore cannot simply place, for example, molten aluminium into contact with a bundle of alumina fibres to produce a composite by infiltration. Also, one cannot sprinkle alumina particles over molten aluminium to see the two phases combine spontaneously into a composite. To overcome the capillary forces that lead to non-wetting, the chemistry of the system must be modified or, alternatively, mechanical force is used (*e.g.*, in infiltration the molten metal can be pushed into the reinforcement preform, in stir-casting an impeller can push the particles into the melt, etc). When mechanical force is used, this energy barrier can be overcome without altering the composite chemical composition.

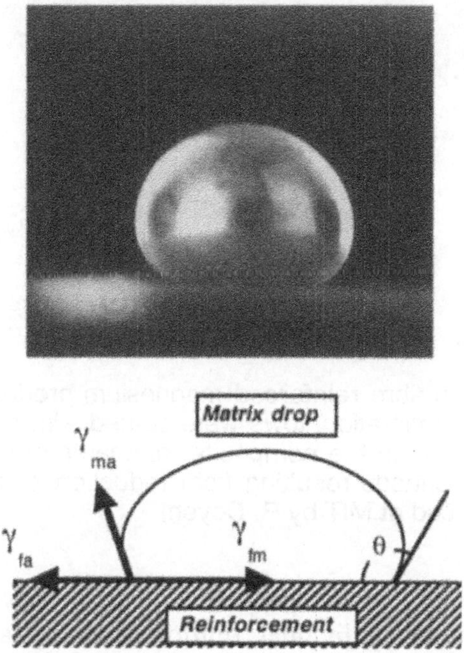

Figure 3.5 - Top: a drop (1 cm wide) of molten Al-4.5 wt pct. Cu at 710°C on alumina in vacuum: poor wetting is apparent by the high contact angle θ, *bottom* (γ_{ma} = matrix surface tension, γ_{fm} = matrix/reinforcement interfacial energy; γ_{fa} = reinforcement surface energy).

Many thermochemical modifications have been proposed to induce favourable wetting of the reinforcement by the matrix. These include processes such as coating the reinforcement (for example with nickel, silica, titanium-boron mixtures, a fluxing agent such as K_2ZrF_6), adding special alloying elements to the matrix, adapting the atmosphere that surrounds the reinforcement when it is placed into contact with the molten matrix metal, and using processing temperatures significantly above the melting point of the metal. Commercial use of chemically activated wetting as in the Lanxide Primex™ process, demonstrates the utility of this approach. Chemical modifications or the use of elevated processing temperatures may, however, lead to undesired phases in the composite as illustrated in Figure 3.6, or may limit the range of compatible matrix alloy chemistries and thus composite properties.

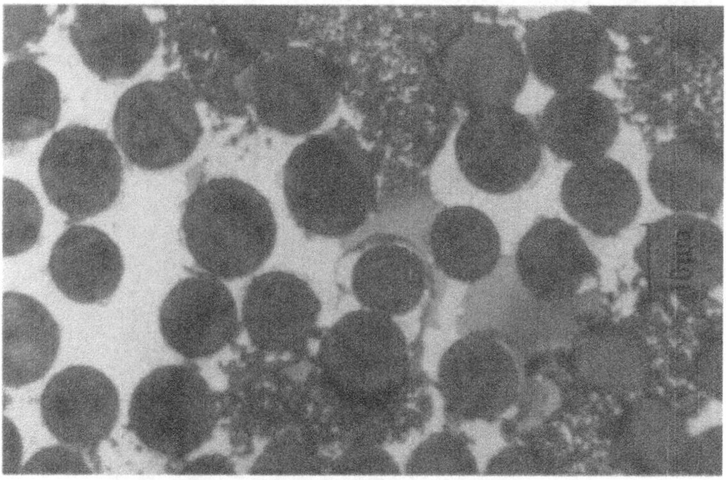

Figure 3.6 - Carbon fibre reinforced magnesium produced by spontaneous infiltration. To drive infiltration, tows were coated with a layer of silica: several second phases result in the composite: remnants of the silica layer, blocky Mg_2Si, dark oxide phases resulting from reduction of the silica by the melt (micrograph produced at MIT by R. Doyen).

Mechanical force can alternatively be used to drive the metal into the open pores between individual reinforcement phase elements, or to introduce the reinforcement into a bath of the matrix. Pressure infiltration is an example: a piston or gas pressurises a molten metal to drive it into the open pores of a preform of the reinforcement as depicted in Figure 3.7. Pressures on the order of 1 MPa (ten atmospheres) are necessary to drive metal into pores 1 μm wide (assuming a typical metal surface tension of 1 J m^2). As it is often

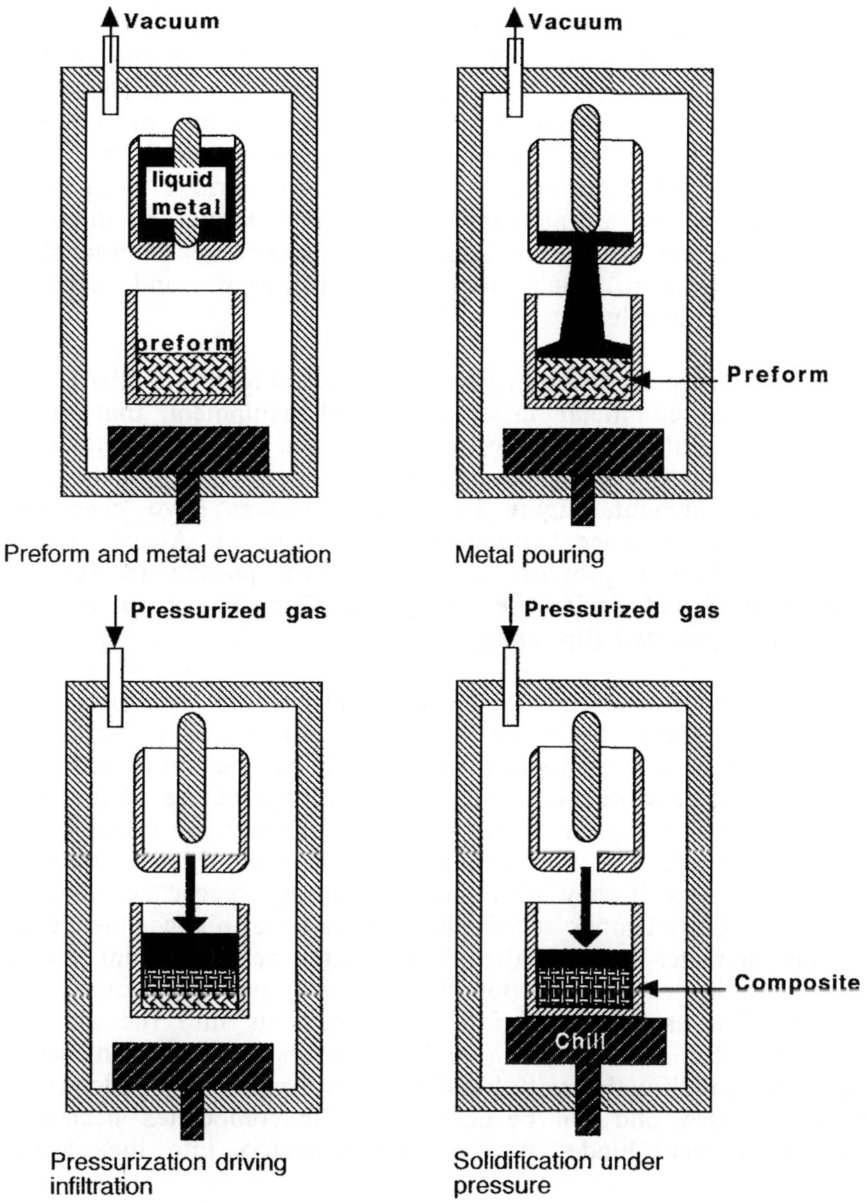

Figure 3.7 - Schematic description of gas-driven pressure infiltration. Several versions exist of this process: for instance, infiltration can also take place upwards, the metal can also be directly placed over the mold before melting, or alternatively can be poured over the preform in a separate chamber than the infiltration pressure chamber.

desirable to infiltrate significantly smaller pores, relatively heavy equipment is necessary to produce and withstand the relatively high pressures that are involved. This is a principal disadvantage of forced wetting processes; however, these processes provide significant latitude for control of the microstructure and optimisation of composite properties. Squeeze casting, pressure die casting (die casting in the US), and pressurized gas infiltration apparatuses of various designs have consequently become a mainstay for the commercial production of MMC. Other forced infiltration techniques have also been demonstrated: ultrasonic infiltration, electromagnetic infiltration, and centrifugal infiltration are examples.

Stir-casting is quite different from infiltration, Figure 3.3. Particles are stirred into molten metal using specialized equipment that forcibly introduces the reinforcement into the liquid metal without simultaneously entraining air. Spray-casting uses inertial forces to combine the matrix and the reinforcement, Figure 3.4. Both processes have been amply demonstrated in practice, largely through work at Alcan, which is probably the largest provider of MMC. The particulate reinforced aluminium marketed under the tradename Duralcan, in particular, is produced by a patented stir-casting process.

After the bond between the liquid metal and the reinforcement is produced, the metal must be solidified. Solidification dictates the microstructure of the composite matrix, and can be influenced by the presence of the reinforcement. Hence, the microstructure of the matrix in the composite will often differ from that of the same metal or alloy solidified analogously in the unreinforced condition. Matrix phases are constrained in size if they grow between closely spaced reinforcement elements that do not move within the composite; as a consequence, their shape and character may be altered. When the reinforcement elements can move in the melt, as is often the case with particles, growing solid metal dendrites can push reinforcement elements into the areas that solidify last (this is seen in Figure 2.4). Feeding of the shrinkage that accompanies solidification with liquid metal is a recurrent challenge in all casting processes, and can be complicated in composites because the reinforcement may hinder flow of molten matrix into the shrinkage cavities.

Interfacial reactions are another recurrent issue in liquid metal processing of metal matrix composites [7,9-11]. At temperatures typical of molten metals, reaction rates can be high. Hence, if the matrix and the reinforcement are thermodynamically prone to react they nearly always will, at least to some extent. A classical example is that of carbon and aluminium, which react to form aluminium carbide, Al_4C_3. This phase grows as discrete platelets embedded in the matrix along the reinforcement surface, and it is weak and hygroscopic. Corrosion properties of the composite are hence degraded, as is the strength of the reinforcement because its surface is prenotched by the reaction phase.

Since interfacial chemical reaction generally leads to the formation of unwanted third phases, and since it may also cause strong degradation of the properties of the reinforcement, it is best avoided. The obvious solution is to use chemically compatible combinations of matrix and reinforcement. Many commercial metal matrix composites are designed with this in mind (*e.g.*, magnesium-free aluminium is compatible with alumina, and aluminium combined with more than about 7 % Si is stable in contact with SiC below 700 °C) [4,10]. Another approach is to minimize the time of contact between matrix and reinforcement at elevated temperatures. This is a key advantage of squeeze-casting and spray-casting processes: they provide short contact times between the matrix and reinforcement at high temperature and rapid solidification of the composite. In comparison, the liquid metal and reinforcement stay in contact much longer during gas-pressure infiltration or stir casting.

b - Solid state processes

Solid state processes can be used to minimize chemical interaction between matrix and reinforcement: both phases remain in the solid state, where diffusivity is much lower. This approach is convenient for the production of titanium matrix composites, because molten titanium (at times called "the universal high-temperature solvent") is very reactive, and degrades nearly any potential reinforcement.

Taking a simple view, solid state composite fabrication requires insertion of solid elements of the matrix between each and every pair of neighbouring reinforcement elements in the composite: in some cases, this can be very difficult. The classical approach is to blend the reinforcement with matrix powder and consolidate the mixture using conventional powder metallurgy processes (cold pressing followed by sintering, hot pressing or isostatic hot pressing, extrusion consolidation, etc.) [2,12-14], as schematically depicted in Figure 3.8. For appropriate blending and densification, the matrix powder must be similar in size or finer than the reinforcement *i.e.*, about 1 to 10 µm in diameter. Such powders are rapidly solidified (given the small size), and exist on the market for several metals used to produce high-performance powder metallurgy alloys or parts. As for unreinforced alloys and parts, the mechanical properties of composites produced by powder metallurgy can benefit from the fine-scale and homogeneous microstructure that rapidly solidified powders yield. These powders are significantly more expensive than bulk metals and alloys, however.

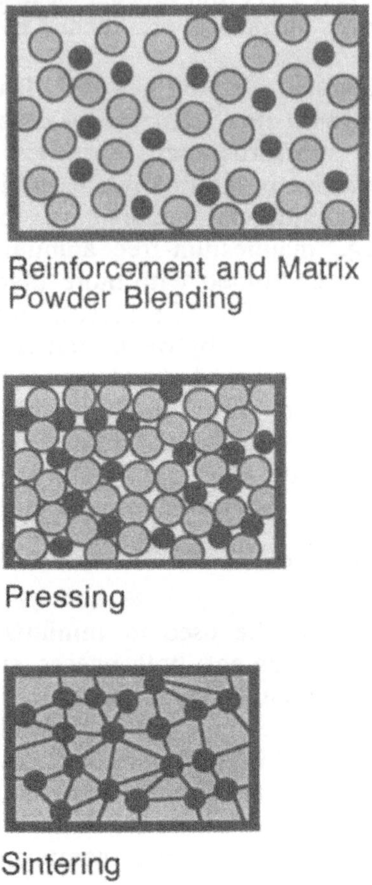

Figure 3.8 - A schematic description of the powder metallurgical fabrication of particulate reinforced metals. Many variants exist of this process: often, pressing and sintering are conducted simultaneously (in hot pressing, or hot isostatic pressing). Also, with oxide-covered metal powders (aluminium in particular), shear often accompanies the densification process to break the oxide layer as it hinders bonding between metal particles.

Conventional powder metallurgical processing is difficult to implement with reinforcements other than equiaxed particles for two basic reasons: (i) blending matrix powder with whiskers or fibres is difficult, and (ii) during consolidation, elongated reinforcements tend to fracture. Powder metallurgical methods have still been used to produce

SiC whisker-reinforced aluminium; however, the whiskers are broken during the process to the extent that the SiC often resembles more a powder in the final composite than the starting whiskers. Conventional solid state powder metallurgical processes are therefore generally used to produce particulate composites.

After the matrix and reinforcement have been blended, Figure 3.9, the mixture is packed into a can, and the can is evacuated and welded shut. This eliminates the presence of gas entrapped in pores within the final composite. During the consolidation step, the pores must be eliminated; this is achieved by cold pressing followed by sintering, or alternatively by hot pressing, by hot isostatic pressing, or using other powder metallurgical processes, all of which require heating. The presence of ceramic reinforcements slows densification of the composite compared to the metal matrix alone, in part because at temperatures appropriate for the sintering of the metal matrix, ceramic reinforcements do not sinter and thus do not contribute to densification.

Figure 3.9 - A partly consolidated hot-pressed blend of TiC and Ti powders: disruptions in the packing density of the Ti powders are apparent. This, together with the fact that the TiC powders do not participate in the sintering process, causes the rate of consolidation of the composite powder blend to be significantly lower than that of the unreinforced matrix (from work reported in Ref. [15]).

For this reason, it is difficult to produce composites with more than about 40 vol % reinforcement using solid-state processes if metal and reinforcement powder particle diameters are of similar orders of magnitude. The extent that the reinforcement hinders densification of the composite depends on the relative sizes of the matrix powder and the reinforcements and on their shapes, thus these parameters also determine the maximum density of reinforcement that can be achieved. In some cases, the oxide layer that coats each matrix powder needs to be broken; this is particularly true of aluminium and magnesium. The composites can then be sheared, either during consolidation (in processes such as extrusion-consolidation), or afterwards during shaping of the densified billet.

One approach to speed densification and to pack particles closer together is to melt the matrix partially. Liquid phase sintering, as this process is termed, is used industrially to produce cemented carbides, and has been used to produce particulate reinforced aluminium composites. Liquid phase sintering has three potential drawbacks in the production of particulate reinforced metals:

(i) because the matrix generally neither wets nor dissolves (partly) the ceramic reinforcements, it is far less efficient that with classical liquid phase sintered materials (as used with cemented carbides or tungsten heavy alloys),
(ii) large intermetallics tend to form in alloy matrices, degrading the properties of the composites; and
(iii) the rates of reaction between the reinforcement and the matrix increase significantly in the presence of a liquid.

For these reasons, liquid phase sintering is seldom used to produce composites such as ceramic particulate reinforced aluminium.

Inserting solid matrix between individual monofilaments is facilitated by their large size. Individual layers of parallel monofilaments and metal foil, powder cloth, or wire can easily be stacked using standard tooling or even by hand and then consolidated by hot pressing, generally after canning and evacuation, Figure 3.10 [16,17]. This basic densification process is often called "diffusion bonding" because it resembles the solid state welding process bearing this name. The matrix and reinforcement must be arranged so as to direct metal flow around the monofilaments and avoid their mutual contact, with a goal to obtain composites with controlled distributions of the reinforcement, free of broken monofilaments and with no porosity. Titanium and titanium aluminide matrix composites reinforced with monofilaments are often produced in this way.

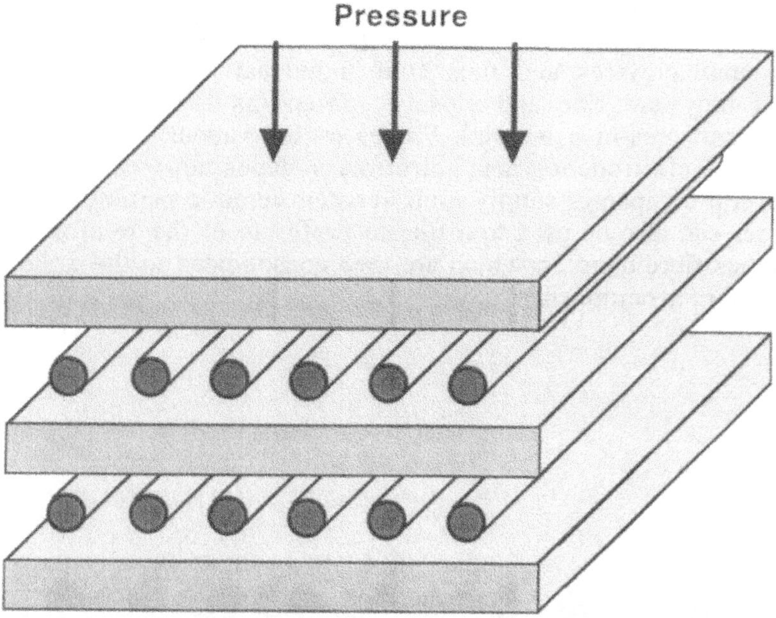

Figure 3.10 - Schematic depiction of diffusion bonding as practiced to consolidate filament reinforced metal matrix composites.

c - Deposition processes

Deposition techniques are used to coat individual reinforcement elements, or a monolayer of reinforcement [16]. These small composite elements are then stacked and consolidated, Figure 3.11. This has the advantage that the deposition step produces the bond between the matrix and the reinforcement and consolidation occurs between layers of metal similar to P/M or other solid metal bonding processes. Deposition processes are also characterized by little if any contact between the reinforcement and molten matrix, which is advantageous for reactive metals such as titanium.

The coating of the reinforcement with metal must be relatively uniform to ensure reproducible stacking and complete densification of the resulting elements. With relatively coarse and regularly spaced reinforcements such as monofilaments, a uniform coating coupled with proper stacking before solid-state consolidation (diffusion bonding) will result in an ideal distribution of reinforcement within in a fully dense matrix, with the option to reach high reinforcement volume fractions.

Methods for coating reinforcement elements include vapour deposition, spray deposition (*e.g.*, plasma-spraying), electrodeposition and electroless deposition. With finer reinforcements, such approaches can also be used, particularly if the deposition process has a natural ability to fill small crevices and thus coat individual reinforcement elements even if they are fine and closely spaced (as is required for 10 μm continuous fibres in a bundle). Processes for vapour phase and electro-deposition (electrodeposition, electroless deposition) can achieve this, and several companies supply metal-coated ceramic particles. Deposition processes can also be used to infiltrate preforms of the reinforcement, or continuous fibre bundles, which are then consolidated in the solid state to produce larger composites.

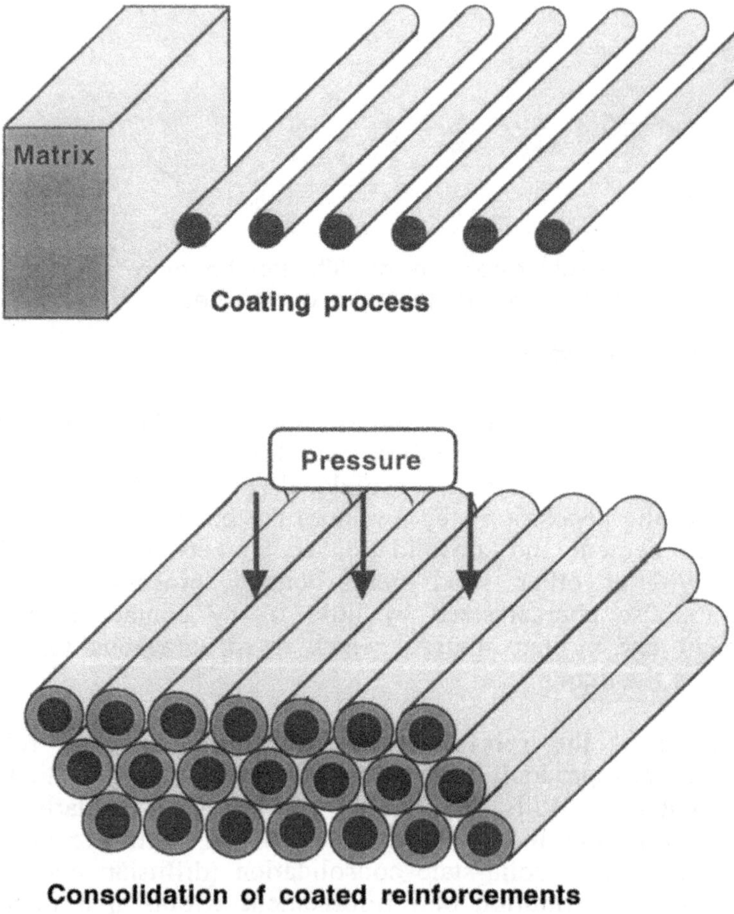

Figure 3.11 - Schematic depiction of monofilament coating and consolidation processing.

d - In-situ processes

In-situ processes produce the matrix and/or the reinforcement while the two phases are combined [18,19]; an example is schematically depicted in Figure 3.12.

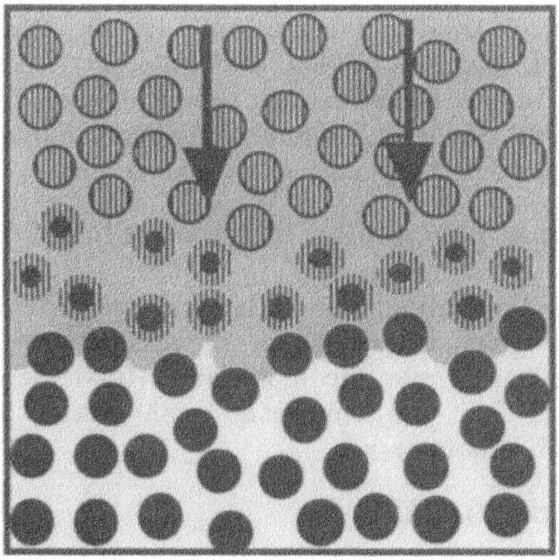

Figure 3.12 - Schematic illustration of one among many *in-situ* processes: reactive infiltration. As the preform (black) is infiltrated by the matrix (light grey), it reacts, either during infiltration (depicted) or afterwards, to form a reinforcement phase (striped), which is a product of chemical reaction between the matrix and the initial preform phase. Many variants of the process exist: for example, the preform can be a mixture of two phases, which interreact through the matrix to form a third, *in-situ* compound reinforcement phase.

The XD™ process is an example of an in-situ process: it employs elementary powder blends of aluminium and two elements that react strongly with one another in molten aluminum to produce a ceramic reinforcement. The result is a composite of a matrix based on aluminium or an aluminide, combined with the *in-situ* synthesized ceramic. For example, Al reinforced with TiB_2 particles can be produced by blending and compacting Al, B and Ti powder particles, and then heating the compact until the aluminium melts. Thereafter, Ti and B react, forming TiB_2. Since the reinforcement (TiB_2) is not initially combined with aluminium but is formed from within the material, the composite is said to be produced via an *in-situ* process. The resulting material is nonetheless a composite, because the titanium diboride was formed by chemical reaction from at least one starting ingredient that always remained distinct from the matrix. This is unlike a conventional two-phase alloy, which was initially fully liquid and in which all solid phases present were either formed by solidification, or by solid-state phase transformations.

There are many other *in-situ* production processes for MMC: some use solids blended in solid or liquids, some use gas, some use a flux. It is, however, worth noting that several materials that have been termed *in-situ* composites in the literature are excluded from the definition used here: these include cast iron, aluminium-silicon alloys and directionally solidified eutectic composites.

3.2 Secondary processing of metal matrix composites

Metal matrix composites may often require further processing after production. The goals of secondary processing are the same as with unreinforced metals: to give the material its final shape, to improve its properties, and to join the composite with other components of the structure in which it is to serve.

a - Net-shape production

The potential for shaping a metal matrix composite depends largely on the nature of the reinforcement. Long continuous fibres or monofilaments, for example, will be severely damaged by any process that imparts significant strain on the material, because more than one or a few percent strain along the reinforcement direction will break it. Significant transverse compression or shear within the plane perpendicular to continuous fibres will also fracture them since they are generally tightly packed in the composite (Figure 2.9). Debonding between matrix and reinforcement may also occur upon straining if the interface is weak, as described in Sections 2.5 and 2.6 for fibres and monofilaments. Significant deformation of infiltrated particulate- and short fibre-reinforced composites may also cause extensive reinforcement fracture by mutual impingement and indentation of contacting tightly packed reinforcement particles or short fibres.

Consequently, long fibre and monofilament reinforced metal matrix composites, as well as high volume fraction particulate reinforced metals and short fibre reinforced metals, are generally produced directly during primary processing to "net shape" or "near-net shape", meaning directly with the shape of the final component save perhaps for some final machining ("near-net shape").

To which extent net-shape composite production is possible depends on the primary manufacturing process and the material. Liquid metal processes such as infiltration can easily produce components of relatively complex shape similar to conventional casting processes. Powder metallurgical processing is also commonly well-suited for net-shape part production; however, if extensive shear is required to promote bonding of oxide-coated metal powder, net-shape powder processes such as hot pressing or hot isostatic pressing may not be appropriate. Processing techniques for monofilament reinforced composites also allow some flexibility in the composite components that can be produced; generally, however, this involves significant labour.

b - Casting

Composites where the particles are surrounded by matrix and do not have a high aspect ratio can accommodate extensive matrix deformation, whether the matrix is liquid or solid. While composite melts are more viscous than molten metal, they can be cast like unreinforced metals using essentially all basic metal casting processes: gravity and pressure die casting, squeeze casting, sand casting and investment mold precision-casting [4]. Some modifications and special care must be exerted, however:

(i) particle segregation due to gravity will occur; the composite melt must therefore be stirred at all times and must solidify before particles can settle in the mould,

(ii) gas bubbles migrate much more slowly through a composite slurry than through unreinforced molten metal and therefore remain within the structure; this is illustrated in Figure 3.13. For this reason, gas entrapment must be carefully avoided while the metal flows, and degassing is more difficult or impossible;

(iii) filtering of metal oxide skins, although possible, is more difficult since particles must not be retained by the filter; this is another reason to avoid turbulence and extensive metal/air contact;

(iv) if matrix and reinforcement react and form an interfacial reaction compound that causes interparticle bonding (such as aluminium carbide in Al/SiC composites), the slurry will become stiff and cease to flow; excessive heating and other factors that may cause the composite to become chemically unstable must therefore be carefully avoided;

(v) chemicals in contact with the composite may cause dewetting of the particles; salt fluxing and bubbling chlorine, for example, produce dewetting in Al/SiC composites (this phenomenon is of course interesting in the context of recycling the material).

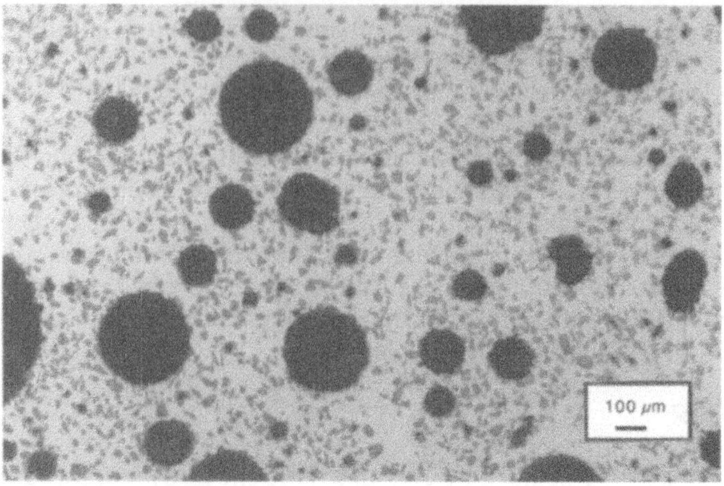

Figure 3.13 - Gas bubbles in SiC reinforced magnesium: bubbles of this magnitude stayed and solidified in the composite due to their very low rate of upward migration, caused by the high viscosity of the composite melt and also by the higher drag induced by particles attached to the bubble. (From research reported in Ref. [20]).

c - Deformation processing

Particulate reinforced metals can also be shaped by conventional solid-state deformation processes, such as extrusion, rolling, and forging [12-14]. Composites are less ductile and more prone to internal damage (in the form of voiding or particle fracture) than unreinforced metals, thus solid-state shaping operations are generally conducted in the hot-working range, *i.e.,* at temperatures sufficiently high that the matrix metal does not work harden. The goal is to minimize reinforcement fracture by lowering the matrix flow stress. Additionally, processes that involve high levels of constraint and of hydrostatic pressure are preferred, such as conventional extrusion, hydrostatic extrusion, and closed-die forging. The reason is that hydrostatic pressure reduces the growth of internal damage whether it evolves as particle fracture, matrix voiding or interfacial decohesion.

Extrusion, in particular hydrostatic extrusion, is thus the most common deformation process for particulate reinforced metal. With tapered dies and appropriate processing parameters, this process provides an efficient means for the production of profiles with nominally equiaxed cross section. Rolling, useful in the production of flat products, is more difficult because edges of the material are unconstrained; given the lower ductility of the composite, this can lead to cracking. Additionally, rolls are cold, introducing temperature gradients across the workpiece, and making process control more difficult than for extrusion. Drawing can also lead to excessive damage build-up within the composite because of the lower hydrostatic pressure in this process compared with extrusion. Excessive tool wear is generally ubiquitous since metal matrix composites typically contain hard ceramic particles that are also used as abrasives.

Besides shaping the composite, solid-state deformation processing generally modifies its microstructure, by distributing the reinforcement more homogeneously and by orienting individual reinforcement elements along the direction of deformation. Tensile ductility, which is particularly sensitive to the reinforcement distribution, can thus be significantly improved by solid-state deformation. Defects can also appear, however, depending on processing conditions: edge cracks in rolling, surface or internal cracks after forging, particle-rich bands in extruded material (these have been attributed to a localization of shear within certain planes during deformation).

Solid-state deformation processes can also be applied to low concentration short fibre and whisker reinforced composites. Deformation processing involving significant reductions (*e.g.*, extrusion ratios in excess of 10), however, will generally result in extensive fracturing of elongated reinforcements. Reinforcement rotation (towards the direction or plane of greatest tensile elongation in the deformation process) on the other hand, becomes quite pronounced for elongated reinforcements such as fibres or whiskers, potentially offering some property enhancements in the direction of alignment. The resulting composite microstructures therefore often feature a well-aligned but much shorter reinforcement, even to the extent that the aspect ratios of the fibrous reinforcements can be reduced to less than five and thus the composites resemble particulate reinforced metals.

The metallurgy of metal matrix composites is somewhat different from that of conventional alloys. Cold-working, a procedure used to harden metals, is often inappropriate for composites, in part because cold deformation risks damaging ceramic reinforcements. In addition, if the diameter of the reinforcement is smaller than about 10 μm, the matrix is cold-worked *in-situ* during cool-down from processing temperature [21-23]. The reason is that, as the composite is cooled, the metal matrix shrinks more than the (ceramic) reinforcement, inducing stresses in the microstructure. Consequently, the reinforcement punches dislocations into the matrix to relieve these thermal stresses, hardening the matrix as if it had been cold-worked. This effect is stronger for smaller reinforcements. Therefore, the mechanical properties of the composites often depend on the size of the reinforcement: the yield strength of the composite is greater for materials reinforced with smaller reinforcements; Figure 3.14 illustrates this effect.

Dislocation densities in composites are therefore typically higher than in similarly processed metals. Consequently, the hardening response of composites with an age-hardening matrix can be changed, particularly if precipitates nucleate and grow on dislocations (see next section). Small amounts of deformation (one or two percent, typical of stretcher-straining) between solution treatment and aging, which change the matrix dislocation density, have been used to alter the age hardening response of particulate reinforced alloys (see next Section) [12].

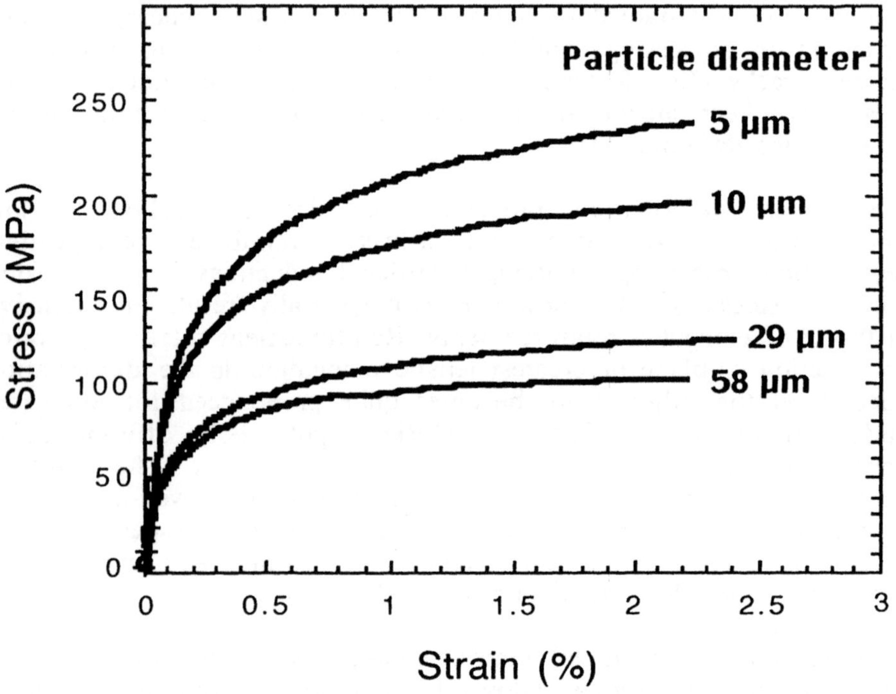

Figure 3.14 - Stress-strain curves of infiltrated alumina particulate reinforced aluminium (nominally 50 vol. %) as a function of particle size: the finer the particle, the harder the matrix and, hence, the greater the composite flow stress. This effect is particularly strong in these composites because of the (comparatively soft) pure aluminium matrix; however, it is apparent in all metal matrix composites. From work reported in Ref. [24].

Discontinuously reinforced metals also display superplastic forming characteristics in a limited range of temperature and deformation rate [12-14]. This feature differs significantly from superplasticity in fine-grained alloys, since superplasticity in composites is exhibited when these are deformed at temperatures very near (or above) the incipient melting temperature of the matrix. Corresponding strain rates are also high (above 10^{-2} s^{-1}), *i.e.*, about two orders of magnitude greater than strain-rates characteristic of "standard" superplastic alloys (10^{-4} s^{-1} - 10^{-3} s^{-1}). Additionally, several studies have demonstrated that large tensile deformations typical of superplasticity can be obtained in particulate reinforced metals by thermal cycling during deformation [14]. On the other hand, reinforcing a superplastic alloy tends to reduce its ability for superplastic deformation.

d - Heat Treating

Metal matrix composites are often produced with a matrix that can be heat-treated, generally by age hardening. Methods, goals and procedures for heat-treating the composite are essentially similar to those used for the matrix; however, there can be an influence of the reinforcement on the ageing process and, therefore, on the appropriate heat-treatment schedule.

If the matrix reacts with the reinforcement, its chemistry may be altered. For example, if the matrix is a 6xxx or 7xxx aluminium alloy, its magnesium content may be reduced by interfacial reaction with reinforcements such as alumina or silica-covered SiC [4,7]. The ageing response of these matrices will then be modified, or sometimes even suppressed. Another effect of the interface in ageing of metal matrix composites can arise if there is accelerated nucleation and growth of hardening precipitates along the interface: this may cause the formation of precipitate-free zones (PFZs) around the reinforcements, analogous to PFZs found along grain boundaries in unreinforced alloys.

As described in the previous section, differential contraction between matrix and reinforcement causes the build-up of internal stresses and an enhanced dislocation density in the matrix of the composite. These factors can also influence ageing kinetics of the matrix. In particular, when hardening precipitates are nucleated on dislocations, ageing kinetics are accelerated in the composite. The time to peak ageing at a given temperature can then be significantly shorter than in the unreinforced matrix processed analogously.

e - Machining

Machining metal matrix composites is difficult: the hard reinforcement, which generally makes these materials wear-resistant, also causes rapid wear of most tooling materials [2,28]. The presence of both a metal and a ceramic can even make these materials "the worst of two worlds". In grinding, for example, composites rich in ceramic tend to require procedures and materials typical of machining bulk ceramics, but the metal "gums up" the tools. In addition, the brittle nature of the reinforcement may lead to significant build-up of damage at the surface of the composite; Figure 3.15 shows an example. Although special machining procedures are required, as a rule metal matrix composites can nonetheless be machined to tight dimensional tolerance and high-quality surface finish.

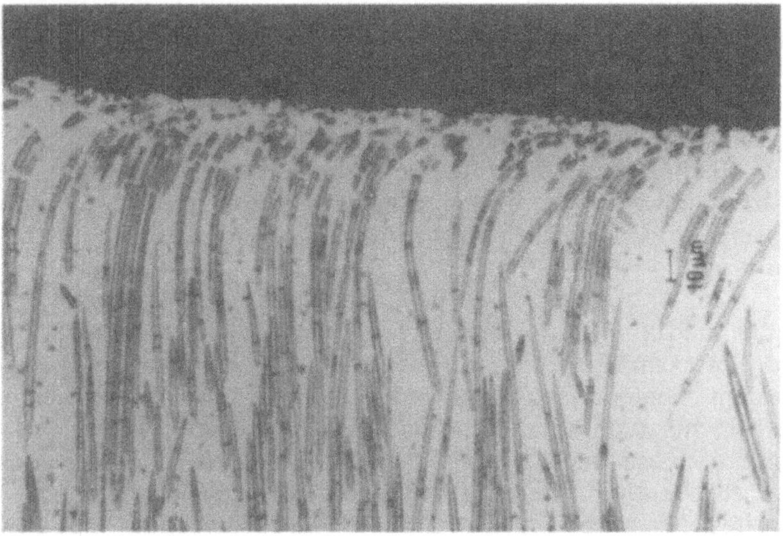

Figure 3.15 - Surface damage caused by machining of an alumina fibre reinforced aluminium composite.

For conventional machining operations (cutting, turning, grinding, etc), diamond-based tools (polycrystalline diamond "PCD" cutting tools and electroplated diamond grinding wheels or hones) are considered superior to all alternatives such as high-speed steel and carbide tool bits. The cost of tool materials for large-scale machining operations is thus greater for metal matrix composites compared with unreinforced alloys. Overall machining costs, however, are not necessarily higher, because diamond tools last long, and permit rapid, aggressive machining. From other standpoints metal matrix composites are attractive with respect to machining: moderate cutting forces and short cutting chips are typical.

As a general trend, the abrasive wear resistance of particulate reinforced metals increases as the size of the reinforcement increases to about 50 μm. This has two important consequences: (i) the ease of machining of composites increases as the reinforcement is made smaller; and (ii) coarser diamond grit in PCD tools lowers tool wear (incidentally, PCD is a metal matrix composite with reinforcements between 10 and 25 μm in size). Some composites have been designed to take advantage of this characteristic; for example, Chesapeake Composites produces composites reinforced with small-diameter particles specifically for ease of machining.

Novel machining techniques have also been tested on metal matrix composites with success. Electro-discharge machining (EDM), which requires a conductive medium, is possible with all but the highest volume fractions and largest reinforcements. Since ceramic reinforcements do not conduct electricity and since only the metal is disintegrated in the process, composites with a continuous or interconnected ceramic reinforcement prove difficult to machine by EDM. Low machining rates, poor surface finish and near-surface damage at higher cutting speeds (which reduces fatigue resistance), are drawbacks of this technique. Still, EDM provides a very attractive method for machining composites reinforced with high concentrations of ceramic (*e.g.*, for infiltrated particulate reinforced metals with about 50 vol. %. ceramic such as the composite in Figure 1.1).

Laser machining has been found to be attractive for high feed rate machining; however, it produces significant thermally-induced damage near the cut surface. Abrasive water jet machining has also been found to be attractive: it is a rapid process that produces no significant heating of the material, thus leaving no heat-affected zone around the cuts. The main drawbacks of abrasive water jet machining are that the machined surface is quite rough, and that the top edges of cuts have a rounded corner.

f - Joining

Since metal matrix composites have a continuous metallic matrix, they can in principle be joined, either with themselves or with unreinforced metals and alloys, using welding and joining techniques conventionally used for their unreinforced matrix [29,30]. This asset has led designers to choose metal matrix composites in applications where weldments are an essential part of the structure, such as automotive driveshafts and bicycle frames.

Composites containing continuous reinforcements pose special challenges, because (i) joining techniques cannot create a continuous bridge of the material across a joint unless the joint runs parallel to the fibres or monofilaments, and (ii) the materials have highly anisotropic properties making joint design more difficult from a mechanical standpoint. On the other hand, weldments along any direction can be made in discontinuously reinforced composites and filler material, when used, can have the same (or similar) composition as the composite. Nonetheless, many of the concerns related to joining MMC are common to all forms of reinforcement.

The material response to the high temperatures associated with many joining techniques is particularly critical for metal matrix composites. If joining exposes the joint and surrounding material to temperatures or chemical environments that induce interfacial reactions between the matrix and reinforcement, the mechanical properties of the joint will at least be changed and will probably be degraded. Since chemical damage of this type is irreversible, control of the joining process (*i.e.*, temperature, time and chemistry) is necessary to avoid reaction. Gas evolution is a second potential source of irreversible degradation in joining composites, and is a common problem in the joining of metals produced by solid-state processes such as powder metallurgy. The highest temperatures and, thus, the greatest risk of degradation in the composite by chemical reaction are with fusion welding processes.

Migration and clustering of the reinforcement in the molten weld are additional concerns in fusion welding processes. Movement of the reinforcement causes heterogeneity in the microstructure of the joint: under the action of gravity, due to convection, due to forces imposed on the reinforcement by the moving solidification front, or (worse) due to dewetting of the particles by the molten matrix. Such migration, clustering, and dewetting will generally weaken and/or embrittle the joint.

Despite these challenges, numerous fusion welding processes have been successfully applied to particulate reinforced metals, including tungsten inert-gas/gas-tungsten arc welding (TIG/GTAW), metal inert-gas/gas-metal arc welding (MIG/GMAW), laser beam welding, electron beam welding, resistance welding, capacitor discharge welding, and plasma joining. High power processes such as laser beam and electron beam welding have the advantage of short times; however, temperatures can be excessively high, thus design of the time-temperature cycles is important to control microstructural reactions and particle migration effects. In contrast, more conventional welding processes, such as TIG and MIG, involve lower temperatures but wider fusion and heat-affected zones, and longer process times. The most appropriate process depends on the material, the joint geometry, and production considerations.

The filler wire is also an important process parameter in fusion welding processes. It can be tailored to avoid problems; for example, interfacial reactivity can be controlled by using a strong carbide former such as Ti to avoid Al_4C_3 formation when welding SiC particulate reinforced aluminium. Reinforcement clumping and dewetting is similarly prevented in TIG welding of Al-Si-Mg alloy/Al_2O_3 composites by employing Mg-rich filler wires. The filler wire itself may be a composite, although composite melts are significantly more viscous than unreinforced molten metals. Consequently mixing of filler and parent metal, as well as flow to fill the joint, may be impaired in fusion welding processes.

When thermal cycles experienced during fusion welding become excessive, lower-temperature solid state welding processes may be attractive. The primary concern in joining MMC by solid-state processes is introducing damage to the reinforcement by excessive deformation, thereby degrading the mechanical properties at the joint. Diffusion bonding, various stir-welding processes, and magnetically impelled arc butt welding have been successfully employed in joining metal matrix composites in the solid state. Lap joints of Ti-6Al-4V reinforced with 40% SiC monofilaments represent a particularly difficult joining task achieved by diffusion bonding.

Transient liquid phase bonding (TLP) is intermediate between liquid and solid state processes and has been used to demonstrate successful bonding in reinforced metals. If a continuously and uniformly reinforced joint is desired, the challenge is to bring the two composites together through diffusion of the bonding layer. Short bonding cycles result in a reinforcement-free bond layer, while longer cycles develop reinforcement clustering along the bond.

Bonding techniques that employ a lower melting point adhesive or interlayer material offer another method for joining composites. These include adhesive bonding, brazing, and soldering. Since bonding temperatures are typically low in comparison with the melting temperature of the matrix, these bonding processes have the advantage that degradation of the composite by interfacial reaction is minimized. Bonding two composite surfaces may be a challenge if the bonding material poorly wets the surface to be bonded; this is a common problem in brazing and soldering because of the inherently poor wetting between most metals and ceramics. This problem may be avoided if the surface of the composite is free of reinforcement, a condition that can be achieved *in-situ* during primary processing of composites by infiltration, for example, or by special secondary surface coating procedures such as aluminising. As in most conventional bonding methods, the bond may substantially change the properties of the component; for example, the presence of a foreign material (the bond layer) may reduce the strength of the bond, particularly at high temperature. The thermal and electrical properties across the bond layer will also be affected, especially if bonding is achieved with polymeric glue.

Joining of a MMC with unreinforced metal can also be achieved at the same time as forming the component. One approach is to produce the composite, and then embed it in a larger metal component by casting for example. This can be challenging in aluminium systems because metals poorly wet the surface oxide that is always present on solid aluminium, causing poor bond strength between the metal and composite. The poor wetting intrinsic to metals on ceramic reinforcements further complicates this problem. Interlayers at the composite surface have been designed to alleviate this difficulty.

Alternatively, the composite and the unreinforced component can be produced in one single processing operation. Infiltration and diffusion bonding are particularly well-suited to a single process that incorporates "selective reinforcement" into a larger metallurgical component. Diesel engine pistons and automotive cylinder blocks with selective reinforcement, for example, are produced by a single casting step where the ceramic preforms are infiltrated at the same time as the larger component is cast. Such selectively reinforced composites feature a metallurgical bond between the composite and unreinforced regions, ideally resulting in excellent properties at the bond. Thermal stresses, however, may degrade the bond strength if the material is not appropriately designed and processed. These stresses are a consequence of the differential shrinkage between the composite and the unreinforced metal, both during solidification and during cooldown to room temperature. Thermal stresses may result in hot-tears along the metal-composite interface that alleviate shrinkage during solidification, or the component may be warped due to the presence of large macro-scale internal stresses.

Chapter 4

METHODOLOGY AND STRUCTURE OF THE SURVEY

4.1 Methodology

The following methods were employed for data collection in building the present survey of the metal matrix composites industry:

(i) all of the information in the database was taken from the public domain;

(ii) while remaining critical and discriminatory in the collection of data, we have trusted published information and we have assumed that companies have provided facts and data in good faith;

(iii) the principal tools that were used in constructing this database have been the Web and direct communication with companies. Every entry in the database is backed by printed documentation kept on file at EPFL. This was done to ensure that the information is traceable, and also to check that it is not confidential. "Word of mouth" or other verbal information, even from trusted colleagues, was discarded if not supported with written confirmation. All information was required to have a reasonably well-defined origin (product bulletins, material data sheets provided by the company or obtained from company web sites, as well as trade journals, academic publications, and written communication from the company providing the material). Web information was printed and stored at the time of use and thus assimilated to printed information on equal footing with documentation provided by the company.

The database was started in 1999. The last comprehensive update of the information was conducted in September 2001, and the final information was added in early November 2001. The database is thus current as of the end of 2001. In January 2003, with the book essentially completed, relevant pages from Chapter 5 of the book were sent to each company, by fax or otherwise by mail (whenever possible). Many responded, and the database was changed where appropriate. Where significant changes were brought to the database, this is indicated at the end of the text describing the company (the database remains, however, dated to the end of 2001).

4.2 Limitations

We have chosen to sacrifice completeness for the sake of rigour: there are certainly (unintended) omissions from the present database. Highly specialised applications, for example from the defence and aerospace industries, are undoubtedly missing. The database also certainly contains errors and flaws; we can only claim that we have done all that we could within our available time and resources to avoid these. We also have sought to be as objective as possible in collecting, collating, and presenting the data; if bias is perceived by the reader it was unintended.

It was not possible to search all possible aspects of the MMC industry, given our resources; thus a number of important limitations exist. The language of the primary source of information is perhaps the foremost limitation. With the assistance of our colleagues, English, French, German, Italian and the Scandinavian languages were all referenced from their primary source. Other European languages needed clarification from the source. The obvious limitation lies in material published exclusively in other languages, particularly those using non-Latin characters, which could not be effectively searched.

This database is not a handbook in the sense of being a definitive collection of carefully collected engineering data on MMC. Rather, it aims to provide a conservative overall description of a relatively new branch of industry, giving only a few basic characteristics of each company, material and product with a primary goal of communicating the current usage and potential of this technology. We have thus not sought to collect extensive design data; it is necessary to consult the material producer for details and validity of material properties. This information should also not be considered as a complete range of available materials neither from the marketplace in general nor the specific producer. This database should therefore not be used to judge or compare companies. It is also worth pointing out that the data provided often represent average expectations for a series of products (properties can vary with the material thickness or diameter, for example).

Since the database emphasises commercial non-secret applications of MMC materials, military applications are often not revealed (except in special cases such as the Title III project for particle or whisker reinforced aluminium material). This is an important limitation of this exercise. As a result, information regarding the MMC activities of the major aircraft and aero-engine manufacturers is sparse and probably not representative of the real situation. In addition, aerospace manufacturers, particularly those based in the US, were generally unwilling to provide any confirmation of their activities (which is perfectly understandable given the interests of the company and the existence of government agreements regarding technology transfer, such as the U.K. National Secrets Act binding companies involved in the defence industry in Great Britain).

4.3 Examples of Inclusion and Omission

Some subjective decision-making was necessary in constructing the database. One example concerns companies that can be considered on the periphery of the MMC industry, namely companies that neither produce MMC for the marketplace nor purchase MMC for integration into products. Companies on the periphery that have a significant interest in the MMC industry were generally included. Tool manufacturers that state their expertise in machining MMC materials and consultant firms that have established track records implementing MMC technology for other companies have, for example, been included. On the other hand, manufacturers of materials used in making composites (fibres, metals, particles, etc) were excluded from the database - except, of course, when they also produce metal matrix composite. Notable exceptions are those companies that work in close relation to the MMC industry, such as manufacturers of short fibre preforms exclusively for liquid-metal infiltration.

For inclusion in this database, it was made a requirement to clearly demonstrate commercial activity of a MMC supplier or, in the case of an application, the existence of the MMC component in a product available in the marketplace. We interpreted demonstration of commercial activity in a restrictive sense: clear evidence of a contract for delivery of a MMC material or a product in the hands of the consumer (whether the general public or a military institution). This often required clarification concerning the level of development and progress towards introduction of the technology into a specific application. When evidence was found of advanced work on processing or an application, the source of the information, the company, or the institute involved in the development was contacted directly to determine if inclusion in the database was justified. For example, published information showed that Volvo had been conducting tests on MMC brake discs manufactured from material produced by Norsk Hydro. Correspondence with both companies failed to confirm if the technology had reached the production stage, hence neither Volvo nor Norsk Hydro were included in the database. Cycotech Pty Ltd., the manufacturer of MMC under the trade name Ultalite, has also developed their material for braking applications in the automotive industry. While Ultalite has yet to be manufactured into brake discs, there is information that a major European automotive manufacturer has placed an order for Ultalite; thus Cyco has been included in the database. These situations are representative of much development work by companies and institutions that are attempting to demonstrate certain performance gains by employing MMC materials. Many of these applications will probably unfold in the future; however, the information obtained in our searches concerning many of these efforts did not clearly demonstrate their tangible present existence in the marketplace.

In a few cases, a company and one of its subsidiaries were both involved with MMC materials, although in a different capacity. For instance, a parent company could be active with the research and development of MMC materials while the subsidiary company was developing industrial applications for the materials. In this case, the parent company was included in the database, provided evidence of marketplace activity exists (the relationship of the companies is then described in the company profile). A similar situation arises with regard to licensing. When a company licenses technology to produce material or a product for a particular world market, both the "inventor" and the "licensee" are included in the database, even if the materials or products are identical. For example, AEA Technology, a British company, licenses its technology to Ametek Specialty Metal Products for the North American market: both companies exist in the database.

The product of this exercise is somewhat ephemeral, a limitation due to the dynamic nature of this young industry. This can be gauged by the number of companies that have withdrawn from MMC activities or declared bankruptcy during the construction of this database. These companies together with their materials and products have been left in the database; however, their names are preceded by the letter "z" (lower-case) so that they appear at the end of the company list when ordered alphabetically. It was deemed desirable to maintain such data since some of these companies have extensive intellectual property concerning MMC materials (and in some cases manufacturing equipment), which may perhaps be available for future licensing or purchase.

In summary, the presentation of the MMC industry that is offered here can be thought of as a broad "lower bound" descriptor of activity in the MMC industry dated to the end of 2001.

4.4 Structure of the database

a - In electronic form

The database was assembled using the FileMaker Pro (version 5) software, available from FileMaker Inc., Santa Clara, CA, USA (mailing address: PO Box 58168, Santa Clara, CA 95052, USA, web site <http://www.filemaker.com/>). It is available on the enclosed disk in the form of four FilemakerPro™ 5 data sets, listing companies, materials, products and references, respectively.

The MMC industry as surveyed is catalogued in the database in the three separate lists of:

(i) companies engaged in commercial MMC business activity,
(ii) MMC materials commercially available, and
(iii) MMC components and in-service products.

Within the database, the data sheets are interlinked: from company in the company data set to its materials in the materials data set, for example. Any of the database files can be opened individually; however, it is generally most convenient to enter through the company or materials data sets.

The *company* data sheets are illustrated and explained in Figure 4.1, as they appear in the electronic database. These datasheets provide, within the database, basic contact information, a summary of manufacturing activity, and mention of special expertise. Companies at all levels of the processing procedure from raw MMC material production to final MMC component are included to the extent possible.

We have sorted companies according to the main steps in metal matrix composite processing. Companies that are engaged in primary processes are called *composite producers,* and often sell bulk metal matrix composite materials. Companies that perform secondary processes to produce semi-finished or finished components, such as particle reinforced aluminium tubing or brake discs, are termed *manufacturers of components*. In many cases, the composite producer and component manufacturer are the same company; for example, net-shape producers of metal matrix composite components are necessarily both. Large producers may also subcontract extrusion or forging operations, for example, and sell semi-finished products. The term *end-user* is used in this context to refer to a company that sells finished products incorporating MMC material, such as a bicycle frame or an automobile or a braking system. Subsidiaries active in the MMC industry are typically not represented by a data sheet, but rather listed on the parent company data sheet in the "other information" section. Special associations such as licensing agreements are also mentioned in this section.

Below this information, the database also gives a short textual description of the company's main features, claims and activities; this text is the same as that which appears in the next chapter of this book.

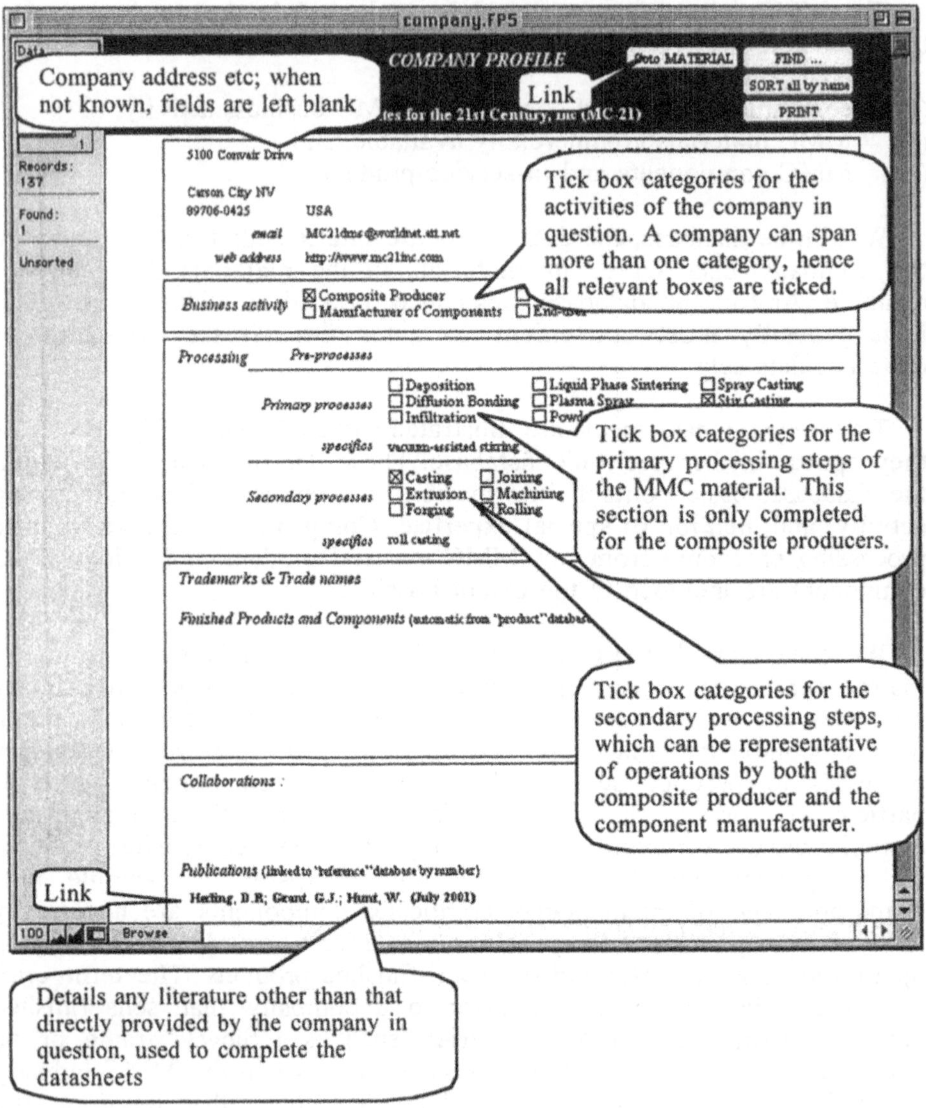

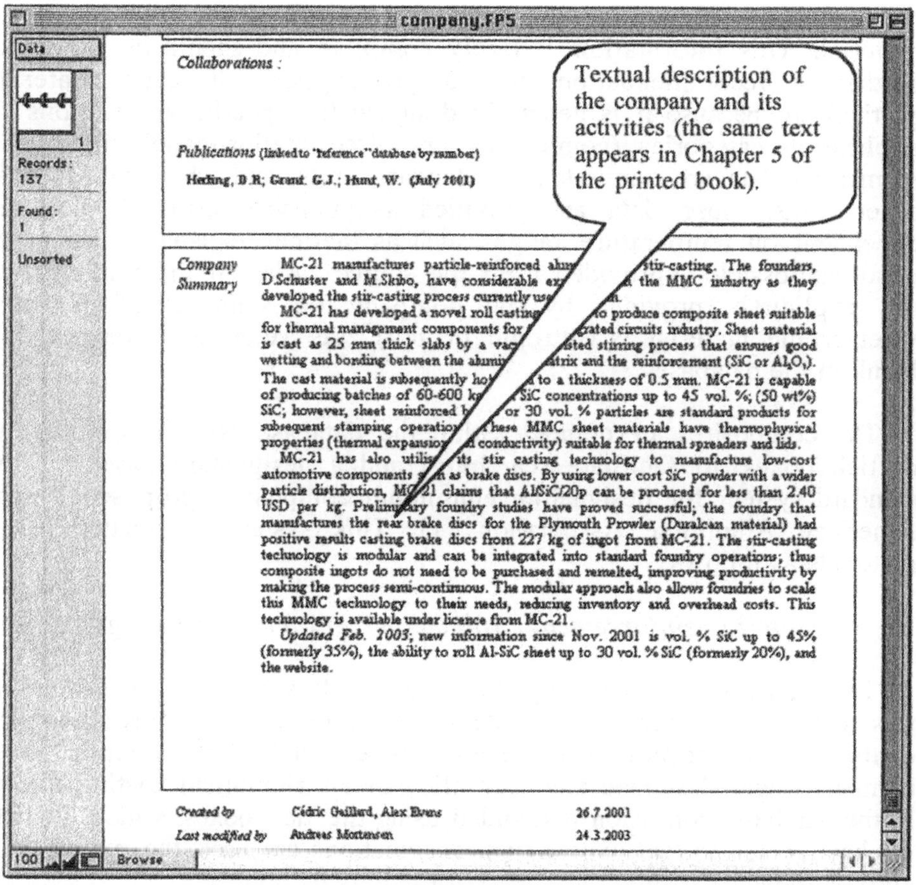

Figure 4.1 (this page and overleaf) - layout, entries and features of Company datasheet in the database.

Companies distinguished by the lower-case letter "z" ahead of their name are, as mentioned above, no longer active in the MMC industry: explanations are given when the details are known.

The *material* data sheets, Figure 4.2 (also depicted as they appear in the database), summarize the general physical characteristics of the material. This information is rarely complete and should be viewed as neither a recommendation nor a proscription of each material, particularly because it is generally designed for specific applications or application categories (composites are often "tailor-made" materials). Numerical data are converted to standard units, and when published in other units, these data are provided as published also. If data are dependent on temperature ranges such as thermal expansion, as much detail as possible is provided from the source. Also, if information was not explicitely provided by the company on some questions (*e.g.*, whether the material is isotropic or not), the relevant boxes are left blank in the database.

The composites are defined within the database using a combination of tick boxes (to facilitate searching) and a designation based on the standard system for designating aluminium composites proposed by the American Aluminum Association [1]. This system designates MMC materials as follows:

matrix / reinforcement / volume percent and morphology

The matrix is referred to by the standard alloy designation. Extending this system to other alloy systems, as done here, requires avoiding confusion with regard to the matrix base metal (for example, three numbers might designate a copper alloy or an aluminium casting alloy). In the database, confusion is avoided using the tick boxes, which identify the base metal. In text, where such designations are used, the matrix base alloy is indicated explicitly when not obvious by context.

The reinforcement is described by its chemical formula or composition and in some cases its brand name, characterizing its special features. The volume percent refers to the reinforcement, and the morphology is an abbreviation for the type of reinforcement: "p" for particle, "sf" for short fibres, "w" for whiskers, "cf" for continuous fibre, or "mf" for monofilaments. Continuous reinforcements ("cf" and "mf") are assumed to be unidirectional unless otherwise specified. Examples are as follows:

201/ C-T300 / 65 cf 0-90°
6061 / Al2O3 / 22 p

In the above CFRM example, 0-90° refers to preforms of fibres that are distributed evenly in perpendicular directions. Additional information is supplied as needed for more specific reference to properties with the following basic format:

... / *form* / *heat treatment*

Form is simply an indicator of final forming step such as "extruded" or "pmc" for permanent mold casting, while the heat treatment is specified as is standard for the matrix alloy such as "T6" or "F" for aluminium matrices. Thus a more specific designation for the second example might be

6061 / Al2O3 / 22 p / extruded / T6

The *product* data sheets, Figure 4.3, briefly describe the MMC components produced commercially and give the companies involved in their production when this information is available. In some cases, products may have been superseded by new designs or other advanced materials or perhaps due to economics; this is indicated as well.

Appendix B provides a list of keywords that can be used to search the database using the tick boxes provided (see Figures 4.1 to 4.3). These search categories are only added for convenience in searching the database for the more habitual processes, materials, properties or applications; they are not exhaustive. For example, the secondary process "rolling" was not included because of the very small number of companies concerned. Still, there is at least one company (MC-21, Metallic Composites for the 21st Century, Inc) who process and deliver particulate reinforced aluminium composites processed by rolling. Such information is often supplied in the textual description of the company and its activities.

Figures 4.1 to 4.3 list some of the features of the electronic database, in the form of "Help" balloons superimposed on sample database sheets as they appear on the screen. The three series of datasheets are interlinked with one another, such that one can switch from one to the other using the toggle buttons provided, or alternatively by clicking on items that appear in bold characters. For instance, clicking on the bold company name appearing on the top of the screen within a materials data sheet toggles the view to the relevant company data sheet. Similarly, clicking on product names appearing in bold on a company data sheet toggles the view towards the relevant product data sheet, and clicking on the "go to materials" button on the company datasheet toggles the view towards the first of the relevant series of materials (sorted by company

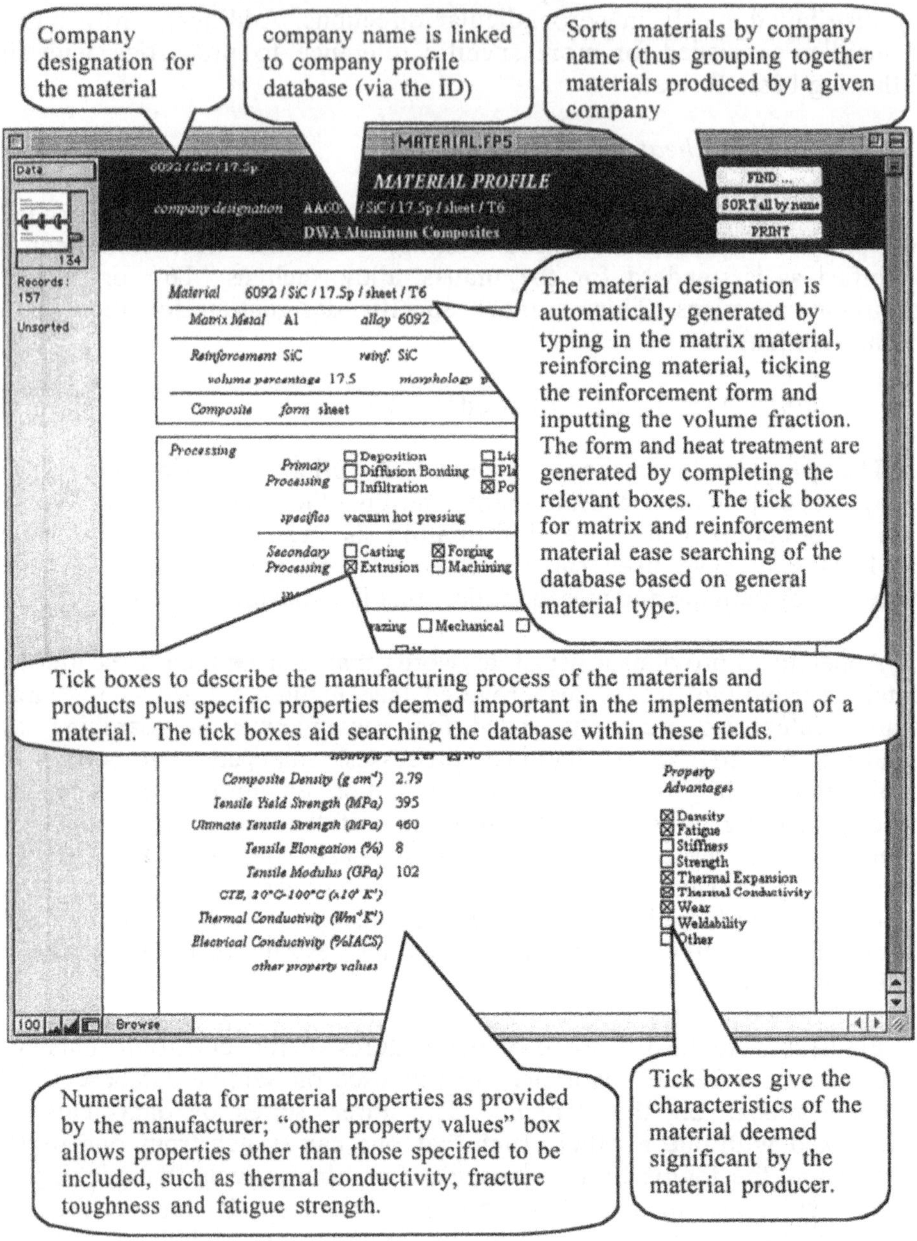
Company designation for the material
company name is linked to company profile database (via the ID)
Sorts materials by company name (thus grouping together materials produced by a given company
MATERIAL.FP5
MATERIAL PROFILE
company designation
DWA Aluminum Composites
FIND ...
SORT all by name
PRINT
Records: 157
Unsorted
Material 6092 / SiC / 17.5p / sheet / T6
Matrix Metal Al
alloy 6092
Reinforcement SiC
reinf. SiC
volume percentage 17.5
Composite form sheet
Processing
Primary Processing
Deposition
Diffusion Bonding
Infiltration
specifics vacuum hot pressing
Secondary Processing
Casting
Forging
Extrusion
Machining
Mechanical
The material designation is automatically generated by typing in the matrix material, reinforcing material, ticking the reinforcement form and inputting the volume fraction. The form and heat treatment are generated by completing the relevant boxes. The tick boxes for matrix and reinforcement material ease searching of the database based on general material type.
Tick boxes to describe the manufacturing process of the materials and products plus specific properties deemed important in the implementation of a material. The tick boxes aid searching the database within these fields.
Composite Density (g cm⁻³) 2.79
Tensile Yield Strength (MPa) 395
Ultimate Tensile Strength (MPa) 460
Tensile Elongation (%) 8
Tensile Modulus (GPa) 102
CTE, 20°C-100°C
Thermal Conductivity
Electrical Conductivity (%IACS)
other property values
Property Advantages
Density
Fatigue
Stiffness
Strength
Thermal Expansion
Thermal Conductivity
Wear
Weldability
Other
Browse
Numerical data for material properties as provided by the manufacturer; "other property values" box allows properties other than those specified to be included, such as thermal conductivity, fracture toughness and fatigue strength.
Tick boxes give the characteristics of the material deemed significant by the material producer.

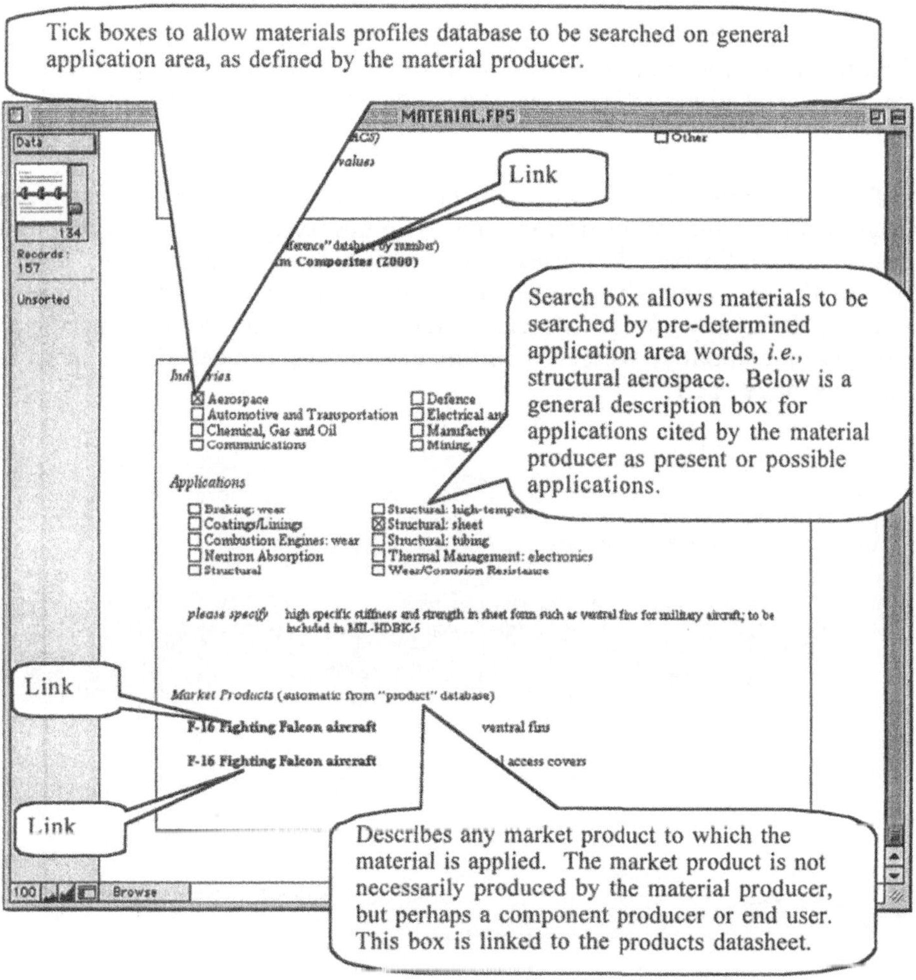

Figure 4.2 (this page and overleaf) - layout, entries and features of Materials datasheet in the database.

name, and hence grouped together) in the materials dataset. A fourth (minor) series of datasheets provides relevant references that have been used in assembling the database; these are also accessible by simple clicking where these appear in bold in other datasheets. These references are also listed alphabetically in the relevant section at the end of this book.

Printing from the databases can be done using the standard Filemaker program command or using the toggle immediately on the database data sheets. To optimize the presentation of data on paper, the printed format differs somewhat from that which appears on the computer screen ("Company" and "Material" data sheets print as two pages per material or company, whereas "Product" and "Reference" data sheets use only one page each).

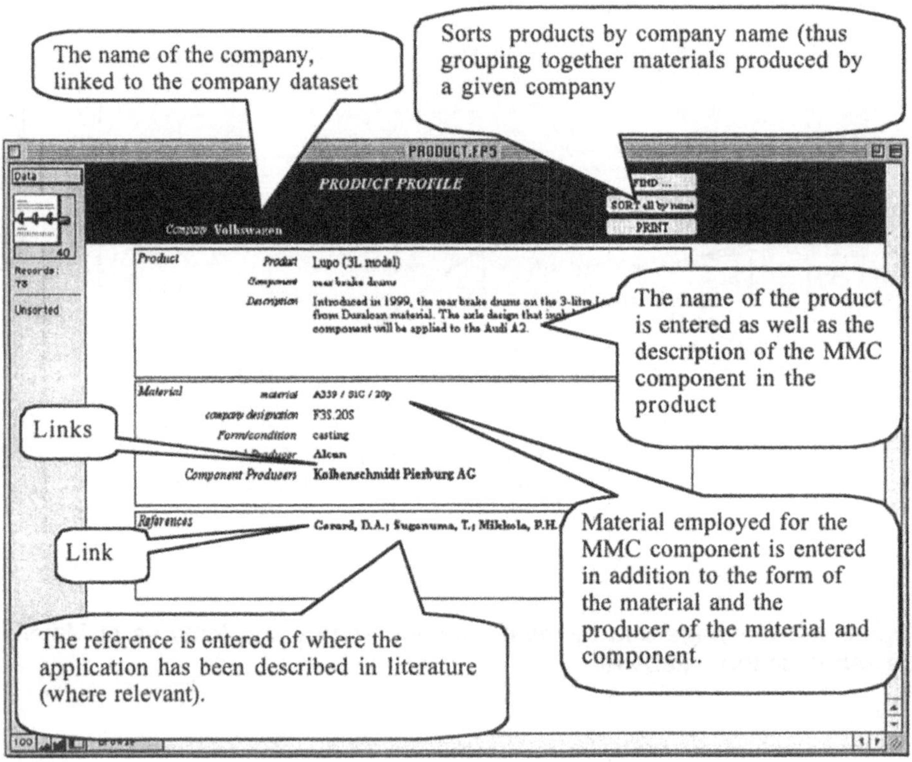

Figure 4.3 - layout, entries and features of Product datasheet in the database.

b - In print

From the electronic database, the following is provided in print within the next chapter, in alphabetical order by company name (with "z-" companies last) and reformatted for improved in-print presentation within the book:

(i) a one-page description for each company, giving its address (to the extent we could document this) and a short textual description of its main features, claims and activities, in alphabetical order by company name (the same text is also in the electronic database and can hence be searched);
(ii) the MMC materials data as given in the database (save for some reformatting), in the form of one-page sheets for each material, located after the relevant company page.

The information provided in this printed version is thus somewhat less exhaustive than in the electronic database. This was done because some information, which is conveniently provided in the form of searchable "tick-box" categories in the electronic database, would be too space-consuming and somewhat redundant in printed form. In particular, the "Products" section of the electronic database is not provided in print in the next chapter, as it would be too space-intensive. Most of the information contained in that portion of the electronic database can, however, be found in the text that describes each company in the next chapter.

Given the far higher efficiency in searching electronically the database, the index provided at the end of the printed book is restricted, for Chapter 5, to coverage of the names of each company entered in the present survey.

Chapter 5

MMC COMPANIES IN ALPHABETICAL ORDER

3M Company, Specialty Fibers and Composites

Metal Matrix Composites
3M Center, Building 207-1W-11
St. Paul MN 55144-1000 USA

tel ++ 1 (651) 733 5168
fax ++ 1 (651) 736-1237
email metalmatrix@mmm.com
web address www.3M.com/mmc

3M has developed a high performance, high strength, metal matrix composite material reinforced with the 3M‘ Nextel‘ Ceramic Oxide Fiber 610. This non-magnetic, low-density material has the strength of steel at a fraction of the weight and has excellent performance at high temperature. Products have been manufactured using traditional aluminum alloys, such as 354, 356, and 6061, and also pure Al and Al-2%Cu. Application and desired performance determine the alloy that is used, as well as the vol. % of Nextel fibres. Common pressure infiltration technologies can be used to manufacture the part, including pressure, squeeze, and die casting. After infiltration, heat treatment can be used to harden the alloy. A final machining is done to ensure precision surfaces.

Several applications have been developed, including pushrods for the racing car industry; these offer significant advantages over steel pushrods including a 40% reduction in weight. The higher MMC stiffness allows the engine to be run at higher speeds, thus providing more power. Brake calipers are another, more recent, development in the automotive sector. The brake caliper is produced by infiltrating a Nextel fibre reinforced insert with aluminum alloy. The insert is used to selectively reinforce areas that experience the highest stresses. The final part is the same size as cast iron, but half the weight. Continuous wire for use in the core of overhead electrical conductors has been undergoing advanced stages of testing with Electricite de France (EDF), the French national utility. This product, which is available from 3M, offers significant advantages over aluminum-steel and all-aluminum cables: low density, high strength, and good electrical conductivity combined with the ability to withstand temperatures of a few hundred degree centigrade. These features address critical needs in congestion on transmission systems while minimizing environmental impact to the surroundings. Current activity focus is in the U.S., including conductor installed at various voltages in Minnesota (in 2001), Hawaii, Tennessee and North Dakota (in 2002). *(information updated in Feb. 2003)*

3M Company, Specialty Fibers and Composites

material (company designation): AMC

Material	Al / Al2O3 / 60cf.		
matrix	unspecified grade Al		
reinforcement	Nextel™ 610 Al2O3	*volume percent*	60
morphology	☐ p ☐ sf ☒ cf ☐ mf ☐ w	*size (μm)*	10-12

Processing	
primary	Infiltration
specifics	pressure infiltration; squeeze casting
secondary	☐ Casting ☐ Forging ☐ Surface Treatments ☐ Extrusion ☐ Machining ☐ Thermal Treatments
specifics	
joining	☐ Brazing ☐ Mechanical ☐ Welding ☐ Other
recyclable	☐ Yes ☐ No

Properties are quoted from manufacturer as general indicators only.

Properties	*isotropic* ☐ Yes ☒ No		*tensile*
density (g cm^{-3})	3.4	*yield strength (MPa)*	1600
CTE (x10^{6} K^{-1})	7 (25-500°C)	*ultimate strength (MPa)*	
thermal cond. (Wm^{-1}K^{-1})		*modulus (GPa)*	240
electrical cond. (%IACS)		*elongation (%)*	

Property Advantages	☒ Density	☒ Strength	☐ Wear
	☒ Fatigue	☒ Thermal Expansion	☐ Weldability
	☒ Stiffness	☐ Thermal Conductivity	☐ Other

Applications	high specific stiffness, strength and fatigue resistance along the fiber direction, electrically conductive, good transverse strength and temperature resistance compared with unidirectional fiber reinforced polymers
Products	pushrods (automotive) brake calipers
Sources	Mendelson, G. (1996). 3M (1999).

3M Company, Specialty Fibers and Composites
material (company designation): AMC

Material Al-2Cu / Al2O3 / 60cf

matrix Al - 2Cu
reinforcement Nextel™ 610 Al2O3 *volume percent* 60
morphology ☐ p ☐ sf ☒ cf ☐ mf ☐ w *size (μm)* 10-12

Processing

primary Infiltration
specifics pressure infiltration; squeeze casting

secondary ☐ Casting ☐ Forging ☐ Surface Treatments ☐ Extrusion ☐ Machining ☐ Thermal Treatments

specifics

joining ☐ Brazing ☐ Mechanical ☐ Welding ☐ Other
recyclable ☐ Yes ☐ No

Properties are quoted from manufacturer as general indicators only.

Properties *isotropic* ☐ Yes ☒ No

			tensile
density (g cm⁻³)	3.4	*yield strength (MPa)*	1500
CTE (x10⁶ K⁻¹)	9 (25-500°C)	*ultimate strength (MPa)*	
thermal cond. (Wm⁻¹K⁻¹)		*modulus (GPa)*	240
electrical cond. (%IACS)		*elongation (%)*	

Property Advantages ☒ Density ☒ Strength ☐ Wear ☒ Fatigue ☐ Thermal Expansion ☐ Weldability ☒ Stiffness ☐ Thermal Conductivity ☐ Other

Applications high specific stiffness, strength and fatigue resistance along the fiber direction, good transverse strength and temperature resistance compared with unidirectional fiber reinforced polymers

Products

Sources Mendelson, G. (1996).
3M (1999).

3M Company, Specialty Fibers and Composites
material (company designation):

Material Al / Al2O3 / 45cf

matrix unspecified grade Al
reinforcement Nextel™ 610 Al2O3 *volume percent* 45
morphology ☐ p ☐ sf ☒ cf ☐ mf ☐ w *size (μm)* 10-12

Processing
primary Infiltration
specifics pressure infiltration; squeeze casting

secondary ☐ Casting ☐ Forging ☐ Surface Treatments ☐ Extrusion ☐ Machining ☐ Thermal Treatments

specifics

joining ☐ Brazing ☐ Mechanical ☐ Welding ☐ Other
recyclable ☐ Yes ☐ No

Properties are quoted from manufacturer as general indicators only.

Properties *isotropic* ☐ Yes ☒ No *tensile*
density ($g\ cm^{-3}$) 3.22 *yield strength (MPa)* 1206
CTE ($x10^6\ K^{-1}$) 6.6 *ultimate strength (MPa)*
thermal cond. ($Wm^{-1}K^{-1}$) *modulus (GPa)* 165
electrical cond. (%IACS) 34.1 *elongation (%)*

Property Advantages ☒ Density ☒ Strength ☐ Wear ☐ Fatigue ☒ Thermal Expansion ☐ Weldability ☐ Stiffness ☒ Thermal Conductivity ☒ Other

Applications high specific stiffness, strength and fatigue resistance along the fiber direction, electrically conductive, good transverse strength and temperature resistance compared with unidirectional fiber reinforced polymers

Products continuous wire High-strength conductor

Sources 3M (1999).

AC Propulsion, Inc

441 Borrego Court

San Dimas CA 91773 USA

tel ++	1 (909) 592 5399
fax ++	1 (909) 394 4598
email	info@acpropulsion.com
web address	www.acpropulsion.com

AC Propulsion manufactures the tzero electric sports car, a high-performance electric-powered vehicle. The lightweight construction of the tzero dictates the use of low-density materials such as MMC materials for brake discs on all four wheels. These MMC brake discs are produced from Duralcan materials from Alcan.

Adtek Intenational Ltd

PO Box 798
Headington
Oxford OX3 0YJ UK

tel ++ 44 (1865) 761 253
fax ++ 44 (1865) 744 221
email bdodd@adtek.u-net.com
web address www.adtek.u-net.com

Adtek develops industrial methods for forming advanced metallic materials such as MMC. One method of plastically forming MMC materials superimposes hydrostatic pressure during deformation processes such as closed and open die forging, extrusion and sheet forming. Adtek has also demonstrated that stepwise deformation with solution treatments between the deformation stages can be used to control cracking during deformation of particle-reinforced aluminium. Adtek obtains MMC materials from Aerospace Metal Composites, specifically aluminium reinforced by silicon carbide particles.

Advanced Metallic Composites

USA

tel ++	1 (440) 275 2755
fax ++	1 (440) 275 1255
email	amcalloy@aol.com
web address	www.advancedmetallic.com

Advanced Metallic Composites designs and manufactures components using MMC materials. Although it has the expertise to implement MMC into a variety of applications, Advanced Metallic Composites is particularly specialised in brake disc technology. To optimise the characteristics of MMC materials in braking applications, Advanced Metallic Composites has developed proprietary material combinations, brake pad materials that work well with MMC rotor materials. The brake discs for the Lotus Elise and Chrysler Prowler were produced by Advanced Metallic Composites from Lanxide materials (Lanxide) and Duralcan materials (Alcan) respectively. The braking technology advanced at Advanced Metallic Composites is not restricted to the automotive industry; brake discs for aircraft, snowmobile and mountain bikes have also been developed.

Advanced Refractory Technologies, Inc (ART)

Suite 310
699 Hertel Avenue
Buffalo NY 14207 USA

tel ++ 1 (716) 875 4091
fax ++ 1 (716) 875 0106
email
web address www.art-inc.com

Advanced Refractory Technology (ART) is developing technology for production of silicon carbide whisker reinforced aluminium for military applications. ART has received over 15 million USD to develop this MMC system for tank track shoes for the Bradley Fighting Vehicle System. Silicon carbide whiskers, manufactured by ART, are formed into preforms by Thermal Ceramics. These preforms are then infiltrated with aluminium alloy A356 using a squeeze casting method performed by General Motors (GM has high volume, near-net shape squeeze casting capability for aluminium).

Replacing steel with whisker-reinforced aluminium track shoes reduces the weight of the track, and extends the lifetime of the component to achieve operations and support savings. This goal is achieved by taking advantage of the enhanced wear resistance of the reinforced-aluminium in selective areas of the component. In addition to the tank track shoe project, whisker reinforced aluminium is being targeted for other military applications, in which there is an operational advantage to replace existing materials. The cost of the lightweight seeker support for AMRAAM (Advanced Medium Range Air-to-Air Missile), for example, can be reduced by replacing the titanium component with reinforced aluminium. A second targeted application of this MMC is to replace the cast iron crown on the high-performance diesel piston for the HMMWV (High Mobility Multipurpose Wheeled Vehicle, the civilian version is called the Hummer).

Tech-Ceram is a division of ART
18 Hunt Rd.
Amesbury MA 01913 USA
tel ++1 (978) 388 2755, fax ++1 (978) 388 6280

Advanced Refractory Technologies, Inc (ART)
material (company designation):

Material A356 / SiC / 20w

matrix Al alloy
reinforcement SiC *volume percent* 20
morphology ☐ p ☐ sf ☐ cf ☐ mf ☒ w *size (μm)* 0.7

Processing
primary Infiltration
specifics squeeze casting by General Motors

secondary ☐ Casting ☐ Forging ☐ Surface Treatments ☐ Extrusion ☐ Machining ☐ Thermal Treatments

specifics

joining ☐ Brazing ☐ Mechanical ☐ Welding ☐ Other
recyclable ☐ Yes ☐ No

Properties are quoted from manufacturer as general indicators only.

Properties *isotropic* ☐ Yes ☐ No *tensile*
density (g cm⁻³) *yield strength (MPa)* 193
CTE (x10⁶ K⁻¹) *ultimate strength (MPa)* 380
thermal cond. (Wm⁻¹K⁻¹) *modulus (GPa)* 96.5
electrical cond. (%IACS) *elongation (%)*

Property Advantages ☒ Density ☒ Strength ☒ Wear ☐ Fatigue ☐ Thermal Expansion ☐ Weldability ☒ Stiffness ☐ Thermal Conductivity ☐ Other

Applications track shoes for the Bradley Fighting Vehicle System; also targeted for lightweight structures in missiles and for high-performance diesel pistons

Products

Sources

AEA Technology Products & Systems
F4 Culham
Abingdon OX14 3DB UK

tel ++ 44 (1235) 463 405
fax ++ 44 (1235) 464 289
email ps-enquiry@aeat.co.uk
web address http://www.aeat.co.uk/

AEA Technology Products and Systems provides innovative engineering solutions to a diverse range of markets including automotive, aerospace, and electronics to name a few. The HIVOL™ family of materials were developed by AEA Products and Systems for thermal management applications. HIVOL™ is aluminium reinforced with silicon carbide, and it can be tailored to suit specific applications, in particular for thermal expansion. This material is produced by pressure-assisted infiltration and can be shaped, machined and plated to meet customer specifications. In North America, HIVOL™ is supplied by Ametek which licences the technology from AEA (materials are listed under Ametek).

AEA Technology Products and Systems is a business of AEA Technology plc.

Update March 2003: the relevant web site for metal matrix composites is
<http://www.aeat.co.uk/prodsys/subdivisions-div/THEM1.html>

Aerospace Metal Composites Limited (AMC)
1 RAE Road
Farnborough, Hampshire GU14 6XE UK

tel ++ 44 (1252) 375 001
fax ++ 44 (1252) 375 002
email david.griffiths@amc-mmc.co.uk
web address www.amc-mmc.co.uk

Aerospace Metal Composites (AMC) supplies both alloys and MMC materials produced by advanced powder metallurgy methods for high-performance applications. AMC produces billet material and has machining capability, in addition to being able to subcontract secondary processing activities.

Two aluminium-matrix composite materials are available from AMC: (i) AMC225xe is 2124 aluminium reinforced with 25 vol. % SiC, available in extrusion, forging or plate; and (ii) AMC640xa is 6061 aluminium reinforced with 40 vol. % SiC, available only as extrusions. AMC aims to offer materials with unsurpassed properties; indeed these materials exhibit high modulus, high strength, and fatigue strength superior to that of high-performance aluminium alloys. Steel-matrix MMC are also being developed at AMC in collaboration with DERA (the UK Defence Evaluation and Research Agency).

MMC brake callipers produced from AMC material have been used for several seasons in Formula 1 racing. Other applications (that are referred to in AMC's literature) are apparently still under development, including aircraft struts and helicopter rotor sleeves, in addition to recreational goods such as golf equipment and bicycle frames. AMC's high-performance MMC are viewed as realistic replacements for carbon fibre-reinforced polymers and for titanium alloys, especially in the aerospace sector where weight and cost reductions are key incentives for materials development that can be garnered by implementing MMC.

In addition to being ISO9002 qualified, AMC complies with Rolls Royce supplier standard CQC103, one of the most demanding benchmarks for quality in the world.

Aerospace Metal Composites Limited (AMC)
material (company designation): AMC225xe

Material 2124 / SiC / 25p / extrusion / T4
matrix Al alloy
reinforcement SiC *volume percent* 25
morphology ☒ p ☐ sf ☐ cf ☐ mf ☐ w *size (μm)*

Processing
primary Powder Metallurgy
specifics

secondary ☐ Casting ☒ Forging ☐ Surface Treatments ☒ Extrusion ☐ Machining ☐ Thermal Treatments
specifics

joining ☐ Brazing ☐ Mechanical ☐ Welding ☐ Other
recyclable ☐ Yes ☐ No

Properties are quoted from manufacturer as general indicators only.

Properties *isotropic* ☐ Yes ☐ No

			tensile
density ($g\ cm^{-3}$)	2.88	*yield strength (MPa)*	480
CTE ($x10^{6}\ K^{-1}$)	15.5 (25°C)	*ultimate strength (MPa)*	700
thermal cond. ($Wm^{-1}K^{-1}$)	150	*modulus (GPa)*	115
electrical cond. (%IACS)		*elongation (%)*	5

Property Advantages ☐ Density ☒ Strength ☐ Wear ☒ Fatigue ☐ Thermal Expansion ☐ Weldability ☒ Stiffness ☐ Thermal Conductivity ☐ Other

Applications

Products

Sources Aerospace Metal Composites Limited (2000).

Aerospace Metal Composites Limited (AMC)
material (company designation): AMC225xe

Material 2124 / SiC / 25p / forging / T4

matrix Al alloy
reinforcement SiC *volume percent* 25
morphology ☒ p ☐ sf ☐ cf ☐ mf ☐ w *size (μm)*

Processing
primary Powder Metallurgy
specifics

secondary ☐ Casting ☒ Forging ☐ Surface Treatments
☒ Extrusion ☐ Machining ☐ Thermal Treatments

specifics

joining ☐ Brazing ☐ Mechanical ☐ Welding ☐ Other
recyclable ☐ Yes ☐ No

Properties are quoted from manufacturer as general indicators only.

Properties *isotropic* ☐ Yes ☐ No *tensile*
density (g cm^{-3}) 2.88 *yield strength (MPa)* 450
CTE (x10^{6} K^{-1}) 15.5 (25°C) *ultimate strength (MPa)* 680
thermal cond. (Wm^{-1}K^{-1}) 150 *modulus (GPa)* 115
electrical cond. (%IACS) *elongation (%)* 5

Property Advantages ☐ Density ☒ Strength ☐ Wear
☒ Fatigue ☐ Thermal Expansion ☐ Weldability
☒ Stiffness ☐ Thermal Conductivity ☐ Other

Applications

Products

Sources Aerospace Metal Composites Limited (2000).

Aerospace Metal Composites Limited (AMC)
material (company designation): AMC225xe

Material 2124 / SiC / 25p / forging / T1

matrix Al alloy

reinforcement SiC *volume percent* 25

morphology ☒ p ☐ sf ☐ cf ☐ mf ☐ w *size (μm)*

Processing

primary Powder Metallurgy

specifics

secondary ☐ Casting ☒ Forging ☐ Surface Treatments ☒ Extrusion ☐ Machining ☐ Thermal Treatments

specifics

joining ☐ Brazing ☐ Mechanical ☐ Welding ☐ Other

recyclable ☐ Yes ☐ No

Properties are quoted from manufacturer as general indicators only.

Properties *isotropic* ☐ Yes ☐ No *tensile*

density (g cm⁻³) 2.88 — *yield strength (MPa)* 300

CTE ($x10^6$ K^{-1}) 15.5 (25°C) — *ultimate strength (MPa)* 470

thermal cond. ($Wm^{-1}K^{-1}$) 150 — *modulus (GPa)* 115

electrical cond. (%IACS) 28 — *elongation (%)* 5

Property Advantages ☐ Density ☒ Strength ☐ Wear ☒ Fatigue ☐ Thermal Expansion ☐ Weldability ☒ Stiffness ☐ Thermal Conductivity ☐ Other

Applications

Products

Sources Aerospace Metal Composites Limited (2000).

Aerospace Metal Composites Limited (AMC)
material (company designation): AMC225xe

Material 2124 / SiC / 25p / plate / T4

matrix Al alloy
reinforcement SiC *volume percent* 25
morphology ☒ p ☐ sf ☐ cf ☐ mf ☐ w *size (μm)*

Processing
primary Powder Metallurgy
specifics

secondary ☐ Casting ☒ Forging ☐ Surface Treatments
☒ Extrusion ☐ Machining ☐ Thermal Treatments

specifics

joining ☐ Brazing ☐ Mechanical ☐ Welding ☐ Other
recyclable ☐ Yes ☐ No

Properties are quoted from manufacturer as general indicators only.

Properties *isotropic* ☐ Yes ☐ No *tensile*
density (g cm^{-3}) 2.88 *yield strength (MPa)* 460
CTE (x10^6 K^{-1}) 15.5 (25°C) *ultimate strength (MPa)* 645
thermal cond. (Wm^{-1}K^{-1}) 150 *modulus (GPa)* 115
electrical cond. (%IACS) *elongation (%)* 3.5

Property Advantages ☐ Density ☒ Strength ☐ Wear
☒ Fatigue ☐ Thermal Expansion ☐ Weldability
☒ Stiffness ☐ Thermal Conductivity ☐ Other

Applications

Products

Sources Aerospace Metal Composites Limited (2000).

Aerospace Metal Composites Limited (AMC)
material (company designation): AMC640xa

Material 6061 / SiC / 40p / extrusion / T6

matrix Al alloy

reinforcement SiC *volume percent* 40

morphology ☒ p ☐ sf ☐ cf ☐ mf ☐ w *size (μm)*

Processing

primary Powder Metallurgy

specifics

secondary ☐ Casting ☒ Forging ☐ Surface Treatments ☒ Extrusion ☐ Machining ☐ Thermal Treatments

specifics

joining ☐ Brazing ☐ Mechanical ☐ Welding ☐ Other

recyclable ☐ Yes ☐ No

Properties are quoted from manufacturer as general indicators only.

Properties *isotropic* ☐ Yes ☐ No

density ($g\ cm^{-3}$)	2.9	*tensile yield strength (MPa)*	480
CTE ($x10^6\ K^{-1}$)	13.4 (25°C)	*ultimate strength (MPa)*	570
thermal cond. ($Wm^{-1}K^{-1}$)	130	*modulus (GPa)*	140
electrical cond. (%IACS)		*elongation (%)*	2.5

Property Advantages ☐ Density ☒ Strength ☐ Wear ☒ Fatigue ☐ Thermal Expansion ☐ Weldability ☒ Stiffness ☐ Thermal Conductivity ☐ Other

Applications

Products

Sources Aerospace Metal Composites Limited (2000).

Aerospace Metal Composites Limited (AMC)
material (company designation): AMC640xa

Material 6061 / SiC / 40p / extrusion / T1

matrix Al alloy
reinforcement SiC *volume percent* 40
morphology ☒ p ☐ sf ☐ cf ☐ mf ☐ w *size (μm)*

Processing
primary Powder Metallurgy
specifics

secondary ☐ Casting ☒ Forging ☐ Surface Treatments ☒ Extrusion ☐ Machining ☐ Thermal Treatments

specifics

joining ☐ Brazing ☐ Mechanical ☐ Welding ☐ Other
recyclable ☐ Yes ☐ No

Properties are quoted from manufacturer as general indicators only.

Properties *isotropic* ☐ Yes ☐ No *tensile*

density ($g\ cm^{-3}$) 2.9
CTE ($x10^6\ K^{-1}$) 13.4 (25°C)
thermal cond. ($Wm^{-1}K^{-1}$) 130
electrical cond. (%IACS) 21

yield strength (MPa) 440
ultimate strength (MPa) 560
modulus (GPa) 140
elongation (%) 3

Property Advantages ☐ Density ☒ Fatigue ☒ Stiffness ☒ Strength ☐ Thermal Expansion ☐ Thermal Conductivity ☐ Wear ☐ Weldability ☐ Other

Applications

Products

Sources Aerospace Metal Composites Limited (2000).

Aerospace Metal Composites Limited (AMC)
material (company designation): Iron MMCs

Material Fe / TiB2 / 15p

matrix Fe
reinforcement TiB2 *volume percent* 15
morphology ☒ p ☐ sf ☐ cf ☐ mf ☐ w *size (μm)*

Processing
primary Powder Metallurgy
specifics

secondary ☐ Casting ☒ Forging ☐ Surface Treatments
☒ Extrusion ☐ Machining ☐ Thermal Treatments
specifics

joining ☐ Brazing ☐ Mechanical ☐ Welding ☐ Other
recyclable ☐ Yes ☐ No

Properties are quoted from manufacturer as general indicators only.

Properties *isotropic* ☐ Yes ☐ No *tensile*
density (g cm^{-3}) 7.36 *yield strength (MPa)* 260
CTE (x10^{6} K^{-1}) *ultimate strength (MPa)* 540
thermal cond. (Wm^{-1}K^{-1}) *modulus (GPa)* 238
electrical cond. (%IACS) *elongation (%)* 20

Property Advantages ☐ Density ☐ Strength ☐ Wear
☐ Fatigue ☐ Thermal Expansion ☐ Weldability
☒ Stiffness ☐ Thermal Conductivity ☐ Other

Applications high stiffness

Products

Sources Aerospace Metal Composites Limited (2000).

Aerospace Metal Composites Limited (AMC)
material (company designation): Iron MMCs

Material Fe / TiB2 / 30p

matrix Fe
reinforcement TiB2 *volume percent* 30
morphology ☒ p ☐ sf ☐ cf ☐ mf ☐ w *size (μm)*

Processing
primary Powder Metallurgy
specifics

secondary ☐ Casting ☒ Forging ☐ Surface Treatments ☒ Extrusion ☐ Machining ☐ Thermal Treatments

specifics

joining ☐ Brazing ☐ Mechanical ☐ Welding ☐ Other
recyclable ☐ Yes ☐ No

Properties are quoted from manufacturer as general indicators only.

Properties *isotropic* ☐ Yes ☐ No *tensile*

density (g cm^{-3}) 6.86 *yield strength (MPa)* 420
CTE (x10^{6} K^{-1}) *ultimate strength (MPa)* 780
thermal cond. (Wm^{-1}K^{-1}) *modulus (GPa)* 270
electrical cond. (%IACS) *elongation (%)* 3

Property Advantages ☐ Density ☐ Strength ☐ Wear ☐ Fatigue ☐ Thermal Expansion ☐ Weldability ☒ Stiffness ☐ Thermal Conductivity ☐ Other

Applications high stiffness

Products

Sources Aerospace Metal Composites Limited (2000).

Aerospace Metal Composites Limited (AMC)
material (company designation): Steel MMCs

Material 316L / TiB_2 / 15p

matrix Fe: stainless steel
reinforcement TiB_2 *volume percent* 15
morphology ☒ p ☐ sf ☐ cf ☐ mf ☐ w *size (μm)*

Processing
primary Powder Metallurgy
specifics

secondary ☐ Casting ☒ Forging ☐ Surface Treatments ☒ Extrusion ☐ Machining ☐ Thermal Treatments
specifics

joining ☐ Brazing ☐ Mechanical ☐ Welding ☐ Other
recyclable ☐ Yes ☐ No

Properties are quoted from manufacturer as general indicators only.

Properties *isotropic* ☐ Yes ☐ No

			tensile
density (g cm^{-3})	7.4	*yield strength (MPa)*	495
CTE (x10^{6} K^{-1})		*ultimate strength (MPa)*	885
thermal cond. (Wm^{-1}K^{-1})		*modulus (GPa)*	218
electrical cond. (%IACS)		*elongation (%)*	11

Property Advantages ☐ Density ☐ Strength ☐ Wear ☐ Fatigue ☐ Thermal Expansion ☐ Weldability ☒ Stiffness ☐ Thermal Conductivity ☒ Other

Applications high stiffness and corrosion resistance

Products

Sources Aerospace Metal Composites Limited (2000).

Aerospace Metal Composites Limited (AMC)
material (company designation): Steel MMCs

Material 316L / TiB2 / 30p

matrix Fe: stainless steel

reinforcement TiB2 *volume percent* 30

morphology ☒ p ☐ sf ☐ cf ☐ mf ☐ w *size (μm)*

Processing

primary Powder Metallurgy

specifics

secondary ☐ Casting ☒ Forging ☐ Surface Treatments ☒ Extrusion ☐ Machining ☐ Thermal Treatments

specifics

joining ☐ Brazing ☐ Mechanical ☐ Welding ☐ Other

recyclable ☐ Yes ☐ No

Properties are quoted from manufacturer as general indicators only.

Properties *isotropic* ☐ Yes ☐ No

density ($g\ cm^{-3}$)	6.9		*tensile*
CTE ($x10^6\ K^{-1}$)		*yield strength (MPa)*	810
thermal cond. ($Wm^{-1}K^{-1}$)		*ultimate strength (MPa)*	1220
electrical cond. (%IACS)		*modulus (GPa)*	250
		elongation (%)	1

Property Advantages ☐ Density ☐ Strength ☐ Wear ☐ Fatigue ☐ Thermal Expansion ☐ Weldability ☒ Stiffness ☐ Thermal Conductivity ☒ Other

Applications high stiffness and corrosion resistance

Products

Sources Aerospace Metal Composites Limited (2000).

Alcan

37676 Enterprise Court

Farmington Hills MI 48331 USA

tel ++ 1 (248) 324 5500
fax ++ 1 (248) 324 5501
email
web address

Alcan is a large multinational company that operates its MMC business through its Global Automotive Products group, indicating the thrust of Alcan's activities with MMC. These materials are manufactured by a patented stir-casting method in batches of 6.8 tonnes. This efficient processing route uses inexpensive raw materials and results in a comparatively low-cost material, in the range of 4 to 11 dollars per kilogram depending on the lot size.

Alcan's MMC are called Duralcan and consist of two basic product ranges: (i) foundry ingots for remelting and casting, for example of brake discs, and (ii) billets for extrusion, particularly of driveshafts. The foundry materials have a silicon carbide reinforcing phase typically in a 300 series aluminium casting alloy. The Duralcan foundry materials are cast by numerous manufacturers especially for braking application; *e.g.*, Hitchiner uses countergravity investment casting to produce high quality MMC castings. The wrought materials are reinforced with alumina and extrusion service providers generally have experience with the Duralcan materials. Apparently, Alcoa extrudes much of the Duralcan tubing. Alcan does not provide extensive technical support for its Duralcan materials, probably because it is not considered a core market. In addition to braking and driveline components, Duralcan is used for tyre studs on the frozen roads of Finland and Scandinavian countries, where winter tyres are a legal requirement. Duralcan 6061/Al2O3/15p material is hot-rolled from billet, drawn into wire, and formed into the tyre studs.

The usage of Duralcan materials is not limited to the automotive industry. The bicycle maker Specialized manufactured a version of its Stumpjumper bicycle from "M2" tubing, Specialized's designation for Duralcan material (6061/Al2O3/10p). This represents a significant application of MMC materials within the recreational sporting market. Other applications of Duralcan materials include metallising wire from AlcoTec; it is also believed that track spikes from Omni-Lite Industries are derived from Duralcan material.

Alcan

material (company designation): F3S.20S

Material A359 / SiC / 20p / pmc / T6

matrix Al alloy

reinforcement SiC *volume percent* 20

morphology ☒ p ☐ sf ☐ cf ☐ mf ☐ w *size (μm)*

Processing

primary Stir Casting

specifics

secondary ☒ Casting ☐ Forging ☐ Surface Treatments ☐ Extrusion ☐ Machining ☒ Thermal Treatments

specifics general purpose gravity casting

joining ☐ Brazing ☐ Mechanical ☐ Welding ☐ Other

recyclable ☒ Yes ☐ No

Properties are quoted from manufacturer as general indicators only.

Properties *isotropic* ☐ Yes ☐ No *tensile*

density ($g\ cm^{-3}$) 2.77 *yield strength (MPa)* 338

CTE ($x10^6\ K^{-1}$) 17.5 (50-100°C) *ultimate strength (MPa)* 359

thermal cond. ($Wm^{-1}K^{-1}$) *modulus (GPa)* 98.6

electrical cond. (%IACS) *elongation (%)* 0.4

Property Advantages ☒ Density ☐ Strength ☐ Wear ☐ Fatigue ☒ Thermal Expansion ☐ Weldability ☒ Stiffness ☐ Thermal Conductivity ☐ Other

Applications general purpose gravity casting for room-temperature: electro-optical castings, brake discs

Products Lupo (3L model) rear brake drums
railway braking system disc brakes

Sources Alcan (1999).

Alcan

material (company designation): F3D.20S

Material A380 / SiC / 20p / pmc / F

matrix Al alloy

reinforcement SiC *volume percent* 20

morphology ☒ p ☐ sf ☐ cf ☐ mf ☐ w *size (μm)*

Processing

primary Stir Casting

specifics

secondary ☒ Casting ☐ Forging ☐ Surface Treatments ☐ Extrusion ☐ Machining ☒ Thermal Treatments

specifics general purpose die casting

joining ☐ Brazing ☐ Mechanical ☐ Welding ☐ Other

recyclable ☒ Yes ☐ No

Properties are quoted from manufacturer as general indicators only.

Properties *isotropic* ☐ Yes ☐ No

density (g cm^{-3})	2.82		*tensile*
CTE (x10^6 K^{-1})	16.9 (50-100°C)	*yield strength (MPa)*	303
thermal cond. (Wm^{-1}K^{-1})		*ultimate strength (MPa)*	352
electrical cond. (%IACS)		*modulus (GPa)*	113.8
		elongation (%)	0.4

Property Advantages ☒ Density ☒ Fatigue ☒ Stiffness ☒ Strength ☐ Thermal Expansion ☐ Thermal Conductivity ☐ Wear ☐ Weldability ☐ Other

Applications general purpose die casting composites: gear box components, friction material back-plath

Products

Sources Alcan (1999).

Alcan

material (company designation): W6A.10A

Material 6061 / Al2O3 / 10p / extruded / T6

matrix Al alloy

reinforcement Al2O3 *volume percent* 10

morphology ☒ p ☐ sf ☐ cf ☐ mf ☐ w *size (μm)*

Processing

primary Stir Casting

specifics

secondary ☐ Casting ☒ Forging ☐ Surface Treatments ☒ Extrusion ☐ Machining ☒ Thermal Treatments

specifics general purpose extrusion

joining ☐ Brazing ☐ Mechanical ☐ Welding ☐ Other

recyclable ☒ Yes ☐ No

Properties are quoted from manufacturer as general indicators only.

Properties *isotropic* ☐ Yes ☐ No *tensile*

density (g cm⁻³) 2.81 *yield strength (MPa)* 300

CTE ($x10^6$ K^{-1}) 37.58 (49-101°C) *ultimate strength (MPa)* 350

thermal cond. ($Wm^{-1}K^{-1}$) *modulus (GPa)* 82

electrical cond. (%IACS) *elongation (%)* 10

Property Advantages ☒ Density ☒ Strength ☐ Wear ☒ Fatigue ☐ Thermal Expansion ☐ Weldability ☒ Stiffness ☐ Thermal Conductivity ☐ Other

Applications general purpose extrusion for room-temperature: bicycle frames; W6S.xx.A (tighter Mg tolerance) for cola-headed tyre studs; W6D.xxA (tighter alloy chemistry) for automotive propeller shafts

Products M2 Stumpjumper bicycle frame

SM2-C lacrosse shaft

Sources Alcan (1999).

Alcan

material (company designation): F3N.20S

Material A360 / SiC / 20p / pmc / F

matrix Al alloy

reinforcement SiC *volume percent* 20

morphology ☒ p ☐ sf ☐ cf ☐ mf ☐ w *size (μm)*

Processing

primary Stir Casting

specifics

secondary ☒ Casting ☐ Forging ☐ Surface Treatments ☐ Extrusion ☐ Machining ☒ Thermal Treatments

specifics die casting

joining ☐ Brazing ☐ Mechanical ☐ Welding ☐ Other

recyclable ☒ Yes ☐ No

Properties are quoted from manufacturer as general indicators only.

Properties *isotropic* ☐ Yes ☐ No

density (g cm^{-3})	2.71	*tensile yield strength (MPa)*	248
CTE (x10^{6} K^{-1})	16.6 (50-100 °C)	*ultimate strength (MPa)*	303
thermal cond. (Wm^{-1}K^{-1})		*modulus (GPa)*	108.2
electrical cond. (%IACS)		*elongation (%)*	0.5

Property Advantages ☒ Density ☒ Strength ☐ Wear ☒ Fatigue ☐ Thermal Expansion ☐ Weldability ☒ Stiffness ☐ Thermal Conductivity ☒ Other

Applications for use in corrosion sensitive applications: contains virtually no Cu and Ni

Products

Sources Alcan (1999).

Alcan

material (company designation): F3S.10S

Material	A359 / SiC / 10p / pmc / T6
matrix	Al alloy
reinforcement	SiC *volume percent* 10
morphology	☒ p ☐ sf ☐ cf ☐ mf ☐ w *size (μm)*

Processing

primary	Stir Casting
specifics	
secondary	☒ Casting ☐ Forging ☐ Surface Treatments ☐ Extrusion ☐ Machining ☒ Thermal Treatments
specifics	general purpose gravity casting
joining	☐ Brazing ☐ Mechanical ☐ Welding ☐ Other
recyclable	☒ Yes ☐ No

Properties are quoted from manufacturer as general indicators only.

Properties   *isotropic* ☐ Yes ☐ No

			tensile
density (g cm^{-3})	2.71	*yield strength (MPa)*	303
CTE (x10^{6} K^{-1})	20.7 (50-100°C)	*ultimate strength (MPa)*	338
thermal cond. (Wm^{-1}K^{-1})		*modulus (GPa)*	86.2
electrical cond. (%IACS)		*elongation (%)*	1.2

Property Advantages ☒ Density ☐ Fatigue ☒ Stiffness ☐ Strength ☒ Thermal Expansion ☐ Thermal Conductivity ☐ Wear ☐ Weldability ☐ Other

Applications general purpose gravity casting for room-temperature: electro-optical castings, brake discs

Products	Galileo™ brake-by-wire	rear brake drums
	EV-1 (electric vehicle)	rear brake drums
Sources	Alcan (1999).	

Alcan

material (company designation): F3D.10S

Material A380 / SiC / 10p / pmc / F

matrix Al alloy

reinforcement SiC *volume percent* 10

morphology ☒ p ☐ sf ☐ cf ☐ mf ☐ w *size (μm)*

Processing

primary Stir Casting

specifics

secondary ☒ Casting ☐ Forging ☐ Surface Treatments ☐ Extrusion ☐ Machining ☒ Thermal Treatments

specifics general purpose die casting

joining ☐ Brazing ☐ Mechanical ☐ Welding ☐ Other

recyclable ☒ Yes ☐ No

Properties are quoted from manufacturer as general indicators only.

Properties *isotropic* ☐ Yes ☐ No

density ($g\ cm^{-3}$)	2.76	*tensile yield strength (MPa)*	241
CTE ($x10^6\ K^{-1}$)	19.3 (50-100°C)	*ultimate strength (MPa)*	345
thermal cond. ($Wm^{-1}K^{-1}$)		*modulus (GPa)*	93.8
electrical cond. (%IACS)		*elongation (%)*	1.2

Property Advantages ☒ Density ☒ Strength ☐ Wear ☒ Fatigue ☐ Thermal Expansion ☐ Weldability ☒ Stiffness ☐ Thermal Conductivity ☐ Other

Applications general purpose die casting composites: gear box components, friction material back-plate

Products

Sources Alcan (1999).

Alcan

material (company designation): W6A.20A

Material 6061 / Al2O3 / 20p / extruded / T6

matrix Al alloy

reinforcement Al2O3 *volume percent* 20

morphology ☒ p ☐ sf ☐ cf ☐ mf ☐ w *size (μm)*

Processing

primary Stir Casting

specifics

secondary ☐ Casting ☒ Forging ☐ Surface Treatments ☒ Extrusion ☐ Machining ☒ Thermal Treatments

specifics general purpose extrusion

joining ☐ Brazing ☐ Mechanical ☐ Welding ☐ Other

recyclable ☒ Yes ☐ No

Properties are quoted from manufacturer as general indicators only.

Properties *isotropic* ☐ Yes ☐ No *tensile*

density (g cm⁻³) 2.86 *yield strength (MPa)* 350

CTE ($x10^6 K^{-1}$) 19.8 (49-100°C) *ultimate strength (MPa)* 370

thermal cond. ($Wm^{-1}K^{-1}$) *modulus (GPa)* 97.7

electrical cond. (%IACS) *elongation (%)* 4

Property Advantages ☒ Density ☒ Strength ☐ Wear ☒ Fatigue ☐ Thermal Expansion ☐ Weldability ☒ Stiffness ☐ Thermal Conductivity ☐ Other

Applications general purpose extrusion for room-temperature: bicycle frames; W6S.xx.A (tighter Mg tolerance) for cola-headed tyre studs; W6D.xxA (tighter alloy chemsitry) for automotive propeller shafts

Products

Sources Alcan (1999).

Alcan

material (company designation): F3N.10S

Material A360 / SiC / 10p / pmc / F

matrix Al alloy

reinforcement SiC *volume percent* 10

morphology ☒ p ☐ sf ☐ cf ☐ mf ☐ w *size (μm)*

Processing

primary Stir Casting

specifics

secondary ☒ Casting ☐ Forging ☐ Surface Treatments ☐ Extrusion ☐ Machining ☒ Thermal Treatments

specifics die casting

joining ☐ Brazing ☐ Mechanical ☐ Welding ☐ Other

recyclable ☒ Yes ☐ No

Properties are quoted from manufacturer as general indicators only.

Properties *isotropic* ☐ Yes ☐ No

density ($g\ cm^{-3}$)	2.65	*tensile yield strength (MPa)*	221
CTE ($x10^6\ K^{-1}$)	21.4 (50-100 °C)	*ultimate strength (MPa)*	310
thermal cond. ($Wm^{-1}K^{-1}$)		*modulus (GPa)*	91
electrical cond. (%IACS)		*elongation (%)*	0.9

Property Advantages ☒ Density ☒ Fatigue ☒ Stiffness ☒ Strength ☐ Thermal Expansion ☐ Thermal Conductivity ☐ Wear ☐ Weldability ☒ Other

Applications for use in corrosion sensitive applications: contains virtually no Cu and Ni

Products

Sources Alcan (1999).

Alcan

material (company designation): W1S.10A

Material 1060 / Al2O3 / 10p

matrix Al alloy

reinforcement Al2O3 *volume percent* 10

morphology ☒ p ☐ sf ☐ cf ☐ mf ☐ w *size (μm)*

Processing

primary Stir Casting

specifics

secondary ☐ Casting ☒ Forging ☐ Surface Treatments ☒ Extrusion ☐ Machining ☐ Thermal Treatments

specifics drawing

joining ☐ Brazing ☐ Mechanical ☐ Welding ☐ Other

recyclable ☒ Yes ☐ No

Properties are quoted from manufacturer as general indicators only.

Properties *isotropic* ☐ Yes ☐ No *tensile*

density (g cm⁻³) → $density\ (g\ cm^{-3})$ *yield strength (MPa)*

CTE ($x10^6\ K^{-1}$) *ultimate strength (MPa)*

thermal cond. ($Wm^{-1}K^{-1}$) *modulus (GPa)*

electrical cond. (%IACS) *elongation (%)*

Property Advantages ☒ Density ☒ Strength ☐ Wear ☒ Fatigue ☐ Thermal Expansion ☐ Weldability ☒ Stiffness ☐ Thermal Conductivity ☐ Other

Applications for metallising wire

Products Duralcan 90/10 metallising wire

Sources Alcan (1999).

Alcan

material (company designation): W2F.20A

Material 2618 / Al2O3 / 20p

matrix Al alloy

reinforcement Al2O3 *volume percent* 20

morphology ☒ p ☐ sf ☐ cf ☐ mf ☐ w *size (μm)*

Processing

primary Stir Casting

specifics

secondary ☐ Casting ☒ Forging ☐ Surface Treatments ☐ Extrusion ☐ Machining ☒ Thermal Treatments

specifics

joining ☐ Brazing ☐ Mechanical ☐ Welding ☐ Other

recyclable ☒ Yes ☐ No

Properties are quoted from manufacturer as general indicators only.

Properties *isotropic* ☐ Yes ☐ No *tensile*

density ($g\ cm^{-3}$) *yield strength (MPa)*

CTE ($x10^6\ K^{-1}$) *ultimate strength (MPa)*

thermal cond. ($Wm^{-1}K^{-1}$) *modulus (GPa)*

electrical cond. (%IACS) *elongation (%)*

Property Advantages ☒ Density ☒ Strength ☐ Wear ☒ Fatigue ☐ Thermal Expansion ☐ Weldability ☒ Stiffness ☐ Thermal Conductivity ☐ Other

Applications for elevated temperature forgings

Products

Sources Alcan (1999).

Alcan

material (company designation): W7A.10A

Material 7005 / Al2O3 / 10p

matrix Al alloy

reinforcement Al2O3 *volume percent* 10

morphology ☒ p ☐ sf ☐ cf ☐ mf ☐ w *size (μm)*

Processing

primary Stir Casting

specifics

secondary ☐ Casting ☒ Forging ☐ Surface Treatments ☒ Extrusion ☐ Machining ☒ Thermal Treatments

specifics

joining ☐ Brazing ☐ Mechanical ☐ Welding ☐ Other

recyclable ☒ Yes ☐ No

Properties are quoted from manufacturer as general indicators only.

Properties *isotropic* ☐ Yes ☐ No *tensile*

density (g cm^{-3}) *yield strength (MPa)*

CTE (x10^{6} K^{-1}) *ultimate strength (MPa)*

thermal cond. (Wm^{-1}K^{-1}) *modulus (GPa)*

electrical cond. (%IACS) *elongation (%)*

Property Advantages ☒ Density ☒ Strength ☐ Wear ☒ Fatigue ☐ Thermal Expansion ☐ Weldability ☒ Stiffness ☐ Thermal Conductivity ☐ Other

Applications high-strength composite

Products

Sources Alcan (1999).

Alcan

material (company designation): F3S.30S

Material A359 / SiC / 30p / F

matrix Al alloy

reinforcement SiC *volume percent* 30

morphology ☒ p ☐ sf ☐ cf ☐ mf ☐ w *size (μm)*

Processing

primary Stir Casting

specifics

secondary ☒ Casting ☐ Forging ☐ Surface Treatments ☐ Extrusion ☐ Machining ☐ Thermal Treatments

specifics

joining ☐ Brazing ☐ Mechanical ☐ Welding ☐ Other

recyclable ☐ Yes ☐ No

Properties are quoted from manufacturer as general indicators only.

Properties *isotropic* ☐ Yes ☐ No

			tensile
density ($g\ cm^{-3}$)	2.80	*yield strength (MPa)*	210
CTE ($x10^{6}\ K^{-1}$)	14.6 (20-100°C)	*ultimate strength (MPa)*	216
thermal cond. ($Wm^{-1}K^{-1}$)		*modulus (GPa)*	120
electrical cond. (%IACS)		*elongation (%)*	1

Property Advantages ☒ Density ☐ Strength ☐ Wear ☐ Fatigue ☒ Thermal Expansion ☐ Weldability ☒ Stiffness ☐ Thermal Conductivity ☐ Other

Applications

Products

Sources

Alcoa	
	201 Isabella Street at the 7th Street Bridge Pittsburgh PA 15212-5858 USA
tel ++	1 (412) 553 4545
fax ++	
email	
web address	www.alcoa.com

Alcoa was a leading MMC producer, but has since withdrawn from the market. Alcoa developed infiltration technology that was sold to Polese. Alcoa also produced particle-reinforced aluminium composites by powder metallurgy for the aerospace industry; for example, Alcoa supplied Forge de Bologne with MMC for forging helicopter blade sleeves for Eurocopter; DWA Aluminum Composites has since become the MMC material supplier for this application. Currently, Alcoa is not publishing MMC related research activities and presumably little if any development work is continuing in this arena.

Alcoa is still active in extruding MMC, particularly the Duralcan materials from Alcan; for example, Alcoa extrudes tubing for driveshaft manufacturers such as American Axle and Manufacturing (AAM) and Visteon.

AlcoTec Wire Corporation
2750 Aero Park Drive
Traverse City MI 49686 USA

tel ++ 1 (231) 941 4111
fax ++ 1 (231) 941 9154
email sales@alcotec.com
web address www.alcotec.com

AlcoTec Wire Corporation is one of the largest manufacturers of aluminium wire for welding and brazing. AlcoTec's extensive inventory of aluminium wire includes Duralcan 90/10, aluminium alloy 1060 reinforced with 10 vol. % (± 2 vol. %) alumina particles (1060 / Al2O3 / 10p). AlcoTec purchases 0.375 inch wire and draws it to smaller diameters. This wire is intended for spray and arc metallising applications in industries that require thermally sprayed coatings for wear and corrosion resistance, and to prevent slipping. Railway hopper cars, for example, have coatings of Duralcan 90/10 applied for wear and corrosion protection.

Update, Feb. 2003: since this database was produced, AlcoTec has exited the Duralcan (9010) market for reasons of availability of the starting composite material in the appropriate form (3/8" rod).

Alexandria Extrusion Company

401 County Road 22 NW

Alexandria MN 56308 USA

tel ++ 1 (320) 763 6537
fax ++ 1 (320) 763 9250
email aecsales@alex-extrusion.com
web address www.alexandriaextrusion.com

Alexandria Extrusion Company (AEC) provides extrusion and machining services for aluminium as well as offering extensive design and prototyping services. AEC has experience extruding aluminium-matrix composites, including "borated aluminium", Al-B composites.

Alloy Technology International, Inc (ATI)
169 Western Highway
West Nyack NY 10994-2637 USA
tel ++ 1 (914) 358 5900 // 1 (800) 431 1854
fax ++ 1 (914) 358 5925
email
web address www.alloytechnology.com

Alloy Technology International (ATI) produces Ferro-Tic®, a family of steel matrix composites with outstanding wear properties. Ferro-Tic® is very hard due to the presence of rounded titanium carbide particles. The carbide also has a low coefficient of friction providing good lubrication and resistance to galling. Other important properties include high temperature tolerance and impact strength. In the annealed state, Ferro-Tic® can be machined by conventional techniques; subsequent heat treatments produce the exceptional hardness and wear properties.

The amount of reinforcing phase can be varied to meet specific requirements. The standard products typically contain between 25 to 45 vol. % titanium carbide. Matrix alloys include medium alloy tools steels, maraging steels and age hardenable nickel alloys. Grade C is recommended for cryogenic applications, and is highly magnetic, replacing A-2 tool steel. Grades CM and CM-25 replace D-2 tool steels for warm metalworking, the latter providing better impact strength and thermal shock resistance. The CHW grades are recommended for hot metalworking to replace H-13 tool steels. To replace S-7 tool steel, ATI offers the SK grade with excellent impact strength. For corrosion resistance of stainless, CS-40 also provides high hardness. Grade PK offers good combination of strength and corrosion resistance. As an alternative to 27-4 PH stainless, Grade MS-5A is FDA approved. The HT-6A grade is a high temperature replacement for Stellite™ and Inconel™.

Ferro-Tic® is supplied in standard round, rings and rectangular flat blanks, although custom materials and sizes are also available. With annual sales between 5-10 million USD, ATI has been successful at marketing the Ferro-Tic® family as attractive replacements for standard tool materials.

Alloy Technology International, Inc (ATI)
material (company designation): Ferro-Tic® Grade C

Material Fe / TiC / 45p

matrix Fe: medium alloy tool steel
reinforcement TiC *volume percent* 45
morphology ☒ p ☐ sf ☐ cf ☐ mf ☐ w *size (μm)*

Processing
primary Powder Metallurgy
specifics

secondary ☐ Casting ☐ Forging ☐ Surface Treatments ☐ Extrusion ☒ Machining ☒ Thermal Treatments
specifics

joining ☐ Brazing ☐ Mechanical ☐ Welding ☐ Other
recyclable ☐ Yes ☐ No

Properties are quoted from manufacturer as general indicators only.

Properties *isotropic* ☐ Yes ☐ No

			tensile
density (g cm^{-3})	6.6	*yield strength (MPa)*	
CTE (x10^{6} K^{-1})	9.772 (21-191°C)	*ultimate strength (MPa)*	
thermal cond. (Wm^{-1}K^{-1})		*modulus (GPa)*	
electrical cond. (%IACS)		*elongation (%)*	

Property Advantages ☐ Density ☐ Strength ☒ Wear ☐ Fatigue ☐ Thermal Expansion ☐ Weldability ☐ Stiffness ☐ Thermal Conductivity ☐ Other

Applications replacement for A-2 tool steel: stamping and piercing operations (also in cryogenic applications); highly magnetic

Products

Sources Alloy Technology International, Inc. (1999).

Alloy Technology International, Inc (ATI)

material (company designation): Ferro-Tic® Grade CM

Material Fe / TiC / 45p

matrix Fe: high chrome tool steel

reinforcement TiC *volume percent* 45

morphology ☒ p ☐ sf ☐ cf ☐ mf ☐ w *size (μm)*

Processing

primary Powder Metallurgy

specifics

secondary ☐ Casting ☐ Forging ☐ Surface Treatments ☐ Extrusion ☒ Machining ☒ Thermal Treatments

specifics

joining ☐ Brazing ☐ Mechanical ☐ Welding ☐ Other

recyclable ☐ Yes ☐ No

Properties are quoted from manufacturer as general indicators only.

Properties *isotropic* ☐ Yes ☐ No *tensile*

density ($g\ cm^{-3}$) 6.45 *yield strength (MPa)*

CTE ($x10^6\ K^{-1}$) 10.27 (21-524°C) *ultimate strength (MPa)*

thermal cond. ($Wm^{-1}K^{-1}$) *modulus (GPa)*

electrical cond. (%IACS) *elongation (%)*

Property Advantages ☐ Density ☐ Strength ☒ Wear ☐ Fatigue ☐ Thermal Expansion ☐ Weldability ☐ Stiffness ☐ Thermal Conductivity ☐ Other

Applications replacement for D-2 tool steel: warm metalworking, drawing and forming (e.g. screws, barrel liners, extrusion molding); magnetic

Products

Sources Alloy Technology International, Inc. (1999).

Alloy Technology International, Inc (ATI)
material (company designation): Ferro-Tic® Grade CM-25

Material Fe / TiC / 25p

matrix Fe: high chrome tool steel

reinforcement TiC *volume percent* 25

morphology ☒ p ☐ sf ☐ cf ☐ mf ☐ w *size (μm)*

Processing

primary Powder Metallurgy

specifics

secondary ☐ Casting ☐ Forging ☐ Surface Treatments ☐ Extrusion ☒ Machining ☒ Thermal Treatments

specifics

joining ☐ Brazing ☐ Mechanical ☐ Welding ☐ Other

recyclable ☐ Yes ☐ No

Properties are quoted from manufacturer as general indicators only.

Properties *isotropic* ☐ Yes ☐ No *tensile*

density $(g\ cm^{-3})$ 7.00 *yield strength (MPa)*

CTE $(x10^{6}\ K^{-1})$ 12.07 (21-538°C) *ultimate strength (MPa)*

thermal cond. $(Wm^{-1}K^{-1})$ *modulus (GPa)*

electrical cond. (%IACS) *elongation (%)*

Property Advantages ☐ Density ☐ Strength ☒ Wear ☐ Fatigue ☐ Thermal Expansion ☐ Weldability ☐ Stiffness ☐ Thermal Conductivity ☐ Other

Applications warm metalworking, drawing and forming (e.g. screws, barrel liners, extrusion molding); highly magnetic

Products

Sources Alloy Technology International, Inc. (1999).

Alloy Technology International, Inc (ATI)
material (company designation): Ferro-Tic® Grade CHW-45

Material Fe / TiC / 45p
matrix Fe: hot work tool steel
reinforcement TiC *volume percent* 45
morphology ☒ p ☐ sf ☐ cf ☐ mf ☐ w *size (μm)*

Processing
primary Powder Metallurgy
specifics

secondary ☐ Casting ☐ Forging ☐ Surface Treatments ☐ Extrusion ☒ Machining ☒ Thermal Treatments
specifics

joining ☐ Brazing ☐ Mechanical ☐ Welding ☐ Other
recyclable ☐ Yes ☐ No

Properties are quoted from manufacturer as general indicators only.

Properties *isotropic* ☐ Yes ☐ No *tensile*
density (g cm^{-3}) 6.45 *yield strength (MPa)*
CTE (x10^{6} K^{-1}) 10.21 (21-538°C) *ultimate strength (MPa)*
thermal cond. (Wm^{-1}K^{-1}) *modulus (GPa)*
electrical cond. (%IACS) *elongation (%)*

Property Advantages ☐ Density ☐ Strength ☒ Wear ☐ Fatigue ☐ Thermal Expansion ☐ Weldability ☐ Stiffness ☐ Thermal Conductivity ☐ Other

Applications replacement for H-13 tool steel: hot metalworking and forming; magnetic

Products

Sources Alloy Technology International, Inc. (1999).

Alloy Technology International, Inc (ATI)

material (company designation): Ferro-Tic® Grade CHW-25

Material Fe / TiC / 25p

matrix Fe: hot work tool steel

reinforcement TiC *volume percent* 25

morphology ☒ p ☐ sf ☐ cf ☐ mf ☐ w *size (μm)*

Processing

primary Powder Metallurgy

specifics

secondary ☐ Casting ☐ Forging ☐ Surface Treatments ☐ Extrusion ☒ Machining ☒ Thermal Treatments

specifics

joining ☐ Brazing ☐ Mechanical ☐ Welding ☐ Other

recyclable ☐ Yes ☐ No

Properties are quoted from manufacturer as general indicators only.

Properties *isotropic* ☐ Yes ☐ No

			tensile
density ($g\ cm^{-3}$)	7.00	*yield strength (MPa)*	
CTE ($x10^6\ K^{-1}$)	11.72 (21-538°C)	*ultimate strength (MPa)*	
thermal cond. ($Wm^{-1}K^{-1}$)		*modulus (GPa)*	
electrical cond. (%IACS)		*elongation (%)*	

Property Advantages ☐ Density ☐ Fatigue ☐ Stiffness ☐ Strength ☐ Thermal Expansion ☐ Thermal Conductivity ☒ Wear ☐ Weldability ☐ Other

Applications replacement for H-13 tool steel: hot extrusion, hot metalworking and forming; highly magnetic

Products

Sources Alloy Technology International, Inc. (1999).

Alloy Technology International, Inc (ATI)

material (company designation): Ferro-Tic® Grade SK

Material Fe / TiC / 35p

matrix Fe: impact resistant tool steel

reinforcement TiC *volume percent* 35

morphology ☒ p ☐ sf ☐ cf ☐ mf ☐ w *size (μm)*

Processing

primary Powder Metallurgy

specifics

secondary ☐ Casting ☐ Forging ☐ Surface Treatments ☐ Extrusion ☒ Machining ☒ Thermal Treatments

specifics

joining ☐ Brazing ☐ Mechanical ☐ Welding ☐ Other

recyclable ☐ Yes ☐ No

Properties are quoted from manufacturer as general indicators only.

Properties *isotropic* ☐ Yes ☐ No *tensile*

density (g cm^{-3}) 6.80 *yield strength (MPa)*

CTE (x10^{6} K^{-1}) 10.59 (21-538 °C) *ultimate strength (MPa)*

thermal cond. (Wm^{-1}K^{-1}) *modulus (GPa)*

electrical cond. (%IACS) *elongation (%)*

Property Advantages ☐ Density ☐ Strength ☒ Wear ☐ Fatigue ☐ Thermal Expansion ☐ Weldability ☐ Stiffness ☐ Thermal Conductivity ☐ Other

Applications replacement for S-7 tool steel; magnetic

Products

Sources Alloy Technology International, Inc. (1999).

Alloy Technology International, Inc (ATI)
material (company designation): Ferro-Tic® Grade CS-40

Material SS / TiC / 45p

matrix Fe: martensitic stainless steel
reinforcement TiC *volume percent* 45
morphology ☒ p ☐ sf ☐ cf ☐ mf ☐ w *size (μm)*

Processing
primary Powder Metallurgy
specifics

secondary ☐ Casting ☐ Forging ☐ Surface Treatments ☐ Extrusion ☒ Machining ☒ Thermal Treatments

specifics

joining ☐ Brazing ☐ Mechanical ☐ Welding ☐ Other
recyclable ☐ Yes ☐ No

Properties are quoted from manufacturer as general indicators only.

Properties *isotropic* ☐ Yes ☐ No *tensile*
density (g cm^{-3}) 6.45 *yield strength (MPa)*
CTE (x10^{6} K^{-1}) 9.16 (21-371°C) *ultimate strength (MPa)*
thermal cond. (Wm^{-1}K^{-1}) *modulus (GPa)*
electrical cond. (%IACS) *elongation (%)*

Property Advantages ☐ Density ☐ Strength ☒ Wear ☐ Fatigue ☐ Thermal Expansion ☐ Weldability ☐ Stiffness ☐ Thermal Conductivity ☒ Other

Applications corrosion resistance of 400 series stainless steel; magnetic

Products

Sources Alloy Technology International, Inc. (1999).

Alloy Technology International, Inc (ATI)

material (company designation): Ferro-Tic® Grade PK

Material Fe / TiC / 42p

matrix Fe: maraging steel

reinforcement TiC *volume percent* 42

morphology ☒ p ☐ sf ☐ cf ☐ mf ☐ w *size (μm)*

Processing

primary Powder Metallurgy

specifics

secondary ☐ Casting ☐ Forging ☐ Surface Treatments ☐ Extrusion ☒ Machining ☒ Thermal Treatments

specifics

joining ☐ Brazing ☐ Mechanical ☐ Welding ☐ Other

recyclable ☐ Yes ☐ No

Properties are quoted from manufacturer as general indicators only.

Properties *isotropic* ☐ Yes ☐ No

			tensile
density ($g\ cm^{-3}$)	6.60	*yield strength (MPa)*	
CTE ($x10^{6}\ K^{-1}$)	10.17 (21-449°C)	*ultimate strength (MPa)*	
thermal cond. ($Wm^{-1}K^{-1}$)		*modulus (GPa)*	
electrical cond. (%IACS)		*elongation (%)*	

Property Advantages ☐ Density ☐ Strength ☒ Wear ☐ Fatigue ☐ Thermal Expansion ☐ Weldability ☐ Stiffness ☐ Thermal Conductivity ☐ Other

Applications magnetic

Products

Sources Alloy Technology International, Inc. (1999).

Alloy Technology International, Inc (ATI)

material (company designation): Ferro-Tic® Grade MS-5A

Material SS / TiC / 41p

matrix Fe: age hardenable martensitic stainless steel

reinforcement TiC *volume percent* 41

morphology ☒ p ☐ sf ☐ cf ☐ mf ☐ w *size (μm)*

Processing

primary Powder Metallurgy

specifics

secondary ☐ Casting ☐ Forging ☐ Surface Treatments ☐ Extrusion ☒ Machining ☒ Thermal Treatments

specifics

joining ☐ Brazing ☐ Mechanical ☐ Welding ☐ Other

recyclable ☐ Yes ☐ No

Properties are quoted from manufacturer as general indicators only.

Properties *isotropic* ☐ Yes ☐ No *tensile*

density ($g\ cm^{-3}$) 6.55 *yield strength (MPa)*

CTE ($x10^6\ K^{-1}$) 9.85 (21-449 °C) *ultimate strength (MPa)* 1358

thermal cond. ($Wm^{-1}K^{-1}$) *modulus (GPa)* 269

electrical cond. (%IACS) *elongation (%)*

Property Advantages ☐ Density ☐ Strength ☒ Wear ☐ Fatigue ☐ Thermal Expansion ☐ Weldability ☐ Stiffness ☐ Thermal Conductivity ☐ Other

Applications replacement for 17-4 PH stainless steel; slightly magnetic; FDA approved

Products

Sources Alloy Technology International, Inc. (1999).

Alloy Technology International, Inc (ATI)

material (company designation): Ferro-Tic® Grade HT-6A

Material Ni / TiC / 40p

matrix Ni: age hardenable nickel base

reinforcement TiC *volume percent* 40

morphology ☒ p ☐ sf ☐ cf ☐ mf ☐ w *size (μm)*

Processing

primary Powder Metallurgy

specifics

secondary ☐ Casting ☐ Forging ☐ Surface Treatments ☐ Extrusion ☒ Machining ☒ Thermal Treatments

specifics

joining ☐ Brazing ☐ Mechanical ☐ Welding ☐ Other

recyclable ☐ Yes ☐ No

Properties are quoted from manufacturer as general indicators only.

Properties *isotropic* ☐ Yes ☐ No *tensile*

density (g cm^{-3}) 6.80 *yield strength (MPa)*

CTE (x10^{6} K^{-1}) 13.21 (21-982 °C) *ultimate strength (MPa)* 1055

thermal cond. (Wm^{-1}K^{-1}) *modulus (GPa)*

electrical cond. (%IACS) *elongation (%)*

Property Advantages ☐ Density ☐ Strength ☒ Wear ☐ Fatigue ☐ Thermal Expansion ☐ Weldability ☐ Stiffness ☐ Thermal Conductivity ☐ Other

Applications replacement for Stellite and Inconel: for wear resistance, also for oxidation and corrosion resistance at high temperature; non-magnetic

Products

Sources Alloy Technology International, Inc. (1999).

Alloytic Co, Ltd

480-7 Gaerang-Ri
Jungnam-Myun
Hwasung-Kun, 445-960 Korea

tel ++ 82 (31) 352 3541
fax ++ 82 (31) 352 3544
email wonrhee@hotmail.com
web address www.alloytic.com

Alloytic produces material of the same name; Alloy-TiC® materials are a family of tool steels reinforced with titanium carbide particles. Alloy-TiC® is available in a variety of grades, all manufactured by a common powder metallurgical method: steel and TiC powders are mixed, then sintered in a vacuum furnace and finally consolidated by hot isostatic pressing (HIPing). In the annealed state, the materials can be machined by conventional methods; after thermal treatments, however, Alloy-TiC® has hardness as high as 70 HRC. This hardness and the presence of about 45 vol. % of smooth, rounded titanium carbide grains gives Alloy-TiC® excellent wear properties. In addition, due to the low density of TiC, Alloy-TiC® materials are of lighter weight than conventional alloys.

Alloy-TiC® has been employed to produce a rod mill guide roller, a component that is exposed to rotation speeds of 35,000 rpm and temperatures of 800-1000°C, in addition to contact with abrasive slag material. Alloy-TiC® rollers perform better compared to conventional tool steels due to their lower weight and improved wear resistance at elevated temperatures.

North American office:
983 Canyon Boulevard
North Vancouver BC
V7R2K2 Canada
tel ++1 (604) 924 8296
fax ++1 (604) 924 8277
email: wonrhee@hotmail.com

Alloytic Co, Ltd

material (company designation): Alloy-TiC FT-1 Grade

Material Fe / TiC / 45p

matrix Fe - 9.7Cr - 3Mo

reinforcement TiC *volume percent* 45

morphology ☒ p ☐ sf ☐ cf ☐ mf ☐ w *size (μm)*

Processing

primary Powder Metallurgy

specifics

secondary ☐ Casting ☐ Forging ☐ Surface Treatments ☐ Extrusion ☒ Machining ☒ Thermal Treatments

specifics

joining ☒ Brazing ☒ Mechanical ☐ Welding ☐ Other

recyclable ☐ Yes ☒ No

Properties are quoted from manufacturer as general indicators only.

Properties *isotropic* ☒ Yes ☐ No *tensile*

density ($g\ cm^{-3}$) 6.5 *yield strength (MPa)*

CTE ($x10^{6}\ K^{-1}$) *ultimate strength (MPa)*

thermal cond. ($Wm^{-1}K^{-1}$) 55 *modulus (GPa)* 303.8

electrical cond. (%IACS) *elongation (%)*

Property Advantages ☒ Density ☐ Strength ☒ Wear ☐ Fatigue ☐ Thermal Expansion ☐ Weldability ☐ Stiffness ☐ Thermal Conductivity ☐ Other

Applications general purpose tooling, heavy forming and warm working; replacement for D-2 tool steel. Product forms include forming rolls, reducing rolls, sizing rolls, guide rollers, drill brushes, polishing discs.

Products

Sources

Alloytic Co, Ltd

material (company designation): Alloy-TiC FT-2 Grade

Material Fe / TiC / 45p

matrix Fe - 18Cr - 0.5Mo

reinforcement TiC *volume percent* 45

morphology ☒ p ☐ sf ☐ cf ☐ mf ☐ w *size (μm)*

Processing

primary Powder Metallurgy

specifics

secondary ☐ Casting ☐ Forging ☐ Surface Treatments ☐ Extrusion ☒ Machining ☒ Thermal Treatments

specifics

joining ☒ Brazing ☒ Mechanical ☐ Welding ☐ Other

recyclable ☐ Yes ☒ No

Properties are quoted from manufacturer as general indicators only.

Properties *isotropic* ☒ Yes ☐ No *tensile*

density ($g\ cm^{-3}$) 6.5

CTE ($x10^{6}\ K^{-1}$)

thermal cond. ($Wm^{-1}K^{-1}$)

electrical cond. (%IACS)

yield strength (MPa)

ultimate strength (MPa)

modulus (GPa) 303.8

elongation (%)

Property Advantages ☒ Density ☐ Strength ☒ Wear ☐ Fatigue ☐ Thermal Expansion ☐ Weldability ☐ Stiffness ☐ Thermal Conductivity ☒ Other

Applications corrosion resistance of 400 series stainless steels: check valves, extruder nipples, knives. Product forms: canning tools, crimp rolls, pelletizing knives, check valves.

Products

Sources

Alloytic Co, Ltd

material (company designation): Alloy-TiC FT-3 Grade

Material Fe / TiC / 35p

matrix Fe - 5.2Cr - 4.1Mo - 0.6Ni

reinforcement TiC *volume percent* 35

morphology ☒ p ☐ sf ☐ cf ☐ mf ☐ w *size (μm)*

Processing

primary Powder Metallurgy

specifics

secondary ☐ Casting ☐ Forging ☐ Surface Treatments ☐ Extrusion ☒ Machining ☒ Thermal Treatments

specifics

joining ☒ Brazing ☒ Mechanical ☐ Welding ☐ Other

recyclable ☐ Yes ☒ No

Properties are quoted from manufacturer as general indicators only.

Properties *isotropic* ☒ Yes ☐ No *tensile*

density ($g\ cm^{-3}$) 6.8 *yield strength (MPa)*

CTE ($x10^6\ K^{-1}$) *ultimate strength (MPa)*

thermal cond. ($Wm^{-1}K^{-1}$) 8.3 *modulus (GPa)* 269.5

electrical cond. (%IACS) *elongation (%)*

Property Advantages ☒ Density ☐ Strength ☒ Wear ☐ Fatigue ☐ Thermal Expansion ☐ Weldability ☐ Stiffness ☐ Thermal Conductivity ☐ Other

Applications replacement for S-7 tool steel; hot work applications, cold heading dies and hammers: hot working rollers, impact dies/punches. Product forms: hot and cold heading die, hot working roller, impacting die/punch.

Products

Sources

Alloytic Co, Ltd

material (company designation): Alloy-TiC FT-4 Grade

Material Fe / TiC / 45p

matrix Fe - 3.1Cr - 3.1Mo

reinforcement TiC *volume percent* 45

morphology ☒ p ☐ sf ☐ cf ☐ mf ☐ w *size (μm)*

Processing

primary Powder Metallurgy

specifics

secondary ☐ Casting ☐ Forging ☐ Surface Treatments ☐ Extrusion ☒ Machining ☒ Thermal Treatments

specifics

joining ☒ Brazing ☒ Mechanical ☐ Welding ☐ Other

recyclable ☐ Yes ☒ No

Properties are quoted from manufacturer as general indicators only.

Properties *isotropic* ☒ Yes ☐ No *tensile*

density ($g\ cm^{-3}$) 6.6 *yield strength (MPa)*

CTE ($x10^6\ K^{-1}$) *ultimate strength (MPa)*

thermal cond. ($Wm^{-1}K^{-1}$) 34.8 *modulus (GPa)* 303.8

electrical cond. (%IACS) *elongation (%)*

Property Advantages ☒ Density ☐ Fatigue ☐ Stiffness ☐ Strength ☐ Thermal Expansion ☐ Thermal Conductivity ☒ Wear ☐ Weldability ☒ Other

Applications replacement for A-2 tool steel: for tools, dies and wear parts such as blanking, stamping, bending dies and punches. Product forms: quills, drawing dies, forming rolls, bending dies and punches.

Products

Sources

Alloytic Co, Ltd

material (company designation): Alloy-TiC FT-5 Grade

Material Ni / TiC / 43p

matrix Ni - 5.7Fe - 12.8Cr - 0.7Al - 0.7Ti

reinforcement TiC *volume percent* 43

morphology ☒ p ☐ sf ☐ cf ☐ mf ☐ w *size (μm)*

Processing

primary Powder Metallurgy

specifics

secondary ☐ Casting ☐ Forging ☐ Surface Treatments ☐ Extrusion ☒ Machining ☒ Thermal Treatments

specifics

joining ☒ Brazing ☒ Mechanical ☐ Welding ☐ Other

recyclable ☐ Yes ☒ No

Properties are quoted from manufacturer as general indicators only.

Properties *isotropic* ☒ Yes ☐ No *tensile*

density (g cm⁻³) 6.8 *yield strength (MPa)*

CTE (x10⁶ K⁻¹) *ultimate strength (MPa)*

thermal cond. (Wm⁻¹K⁻¹) *modulus (GPa)* 309.7

electrical cond. (%IACS) *elongation (%)*

Property Advantages ☒ Density ☐ Strength ☒ Wear ☐ Fatigue ☐ Thermal Expansion ☐ Weldability ☐ Stiffness ☐ Thermal Conductivity ☒ Other

Applications replacement for Stellite and Inconel: for high-temperature wear applications such as hot forging die inserts, hot pressing dies and punches; non-magnetic. Product forms: copper extrusion dies, hot forging die inserts.

Products

Sources

Alloytic Co, Ltd

material (company designation): Alloy-TiC FT-7 Grade

Material Fe / TiC / 44p

matrix Fe - 9.4Cr - 3.5Mo - 0.7Al - 2.8Ni - 0.7Ti

reinforcement TiC *volume percent* 44

morphology ☒ p ☐ sf ☐ cf ☐ mf ☐ w *size (μm)*

Processing

primary Powder Metallurgy

specifics

secondary ☐ Casting ☐ Forging ☐ Surface Treatments ☐ Extrusion ☒ Machining ☒ Thermal Treatments

specifics

joining ☒ Brazing ☒ Mechanical ☐ Welding ☐ Other

recyclable ☐ Yes ☒ No

Properties are quoted from manufacturer as general indicators only.

Properties *isotropic* ☒ Yes ☐ No *tensile*

density (g cm^{-3}) 6.6 *yield strength (MPa)*

CTE (x10^{6} K^{-1}) *ultimate strength (MPa)*

thermal cond. (Wm^{-1}K^{-1}) *modulus (GPa)* 289.1

electrical cond. (%IACS) *elongation (%)*

Property Advantages ☒ Density ☐ Strength ☒ Wear ☐ Fatigue ☐ Thermal Expansion ☐ Weldability ☐ Stiffness ☐ Thermal Conductivity ☒ Other

Applications replacement for 17-4 PH stainless: for food and chemical industries and processing plastics; slightly magnetic. Product forms: pelletizing knives and dies, injection moulding nozzles, face seals.

Products

Sources

AlumiPlate, Inc

8960 Springbrook Drive

Minneapolis MN 55433-5874 USA

tel ++ 1 (612) 786 3788
fax ++ 1 (612) 786 8518
email info@alumiplate.com
web address www.alumiplate.com

AlumiPlate offers metal finishing services for a variety of applications including the electronic packaging industry. The AlumiPlate™ technology enables plating components with aluminium to provide a dense surface of highly corrosion-resistant pure aluminium. The aluminium surface can be subsequently anodized to form a very hard, dielectric surface without affecting the thermal conductivity of the bulk. The anodised surface protects the component from corrosion and scoring as well as providing an electrically insulating layer. To increase absorptivity and emissivity, this surface can also be dyed black. In addition, the plating can provide a surface that is suitable for vacuum brazing with other materials, as often necessary when creating electronic packages.

The AlumiPlate™ technology is ideal for thermal management solutions with MMC (e.g. the Al-SiC family of materials), since these composites tend to be difficult to braze and anodise, and may not exhibit the necessary surface characteristics such as emissivity.

American Axle and Manufacturing, Inc (AAM)

Sales and Marketing
2965 Technology Drive
Rochester Hills MI 48309 USA

tel ++ 1 (800) 299 2953
fax ++
email sales@aam.com
web address www.aam.com

American Axle and Manufacturing (AAM) is a global supplier of driveline systems, chassis systems, and forged products for automotive applications. AAM complements its complete line of driveline products with the benefits of a particle-reinforced aluminium rear propeller shaft. The one-piece design results in a lightweight solution and eliminates the need for an intermediate joint and support bracket. The high specific stiffness of particle-reinforced aluminium leads to higher rotational speeds and reduced inertia. In addition, aluminium-based propeller shafts are corrosion resistant, a benefit compared to steel shafts. MMC shafts are available in all standard configurations and sizes, and also available in greater lengths than steel shafts. Alcoa extrudes Duralcan material from Alcan for AAM.

American Eagle Mountain Bikes

USA

tel ++
fax ++
email info@americaneagle-bikes.com
web address www.americaneagle-bikes.com

American Eagle manufactures the Lightning 890 mountain bike using a MMC frame.

American Roller Company

2223 Lakeside Drive

Bannockburn IL 60015 USA

tel ++ 1 (847) 295 6750
fax ++ 1 (847) 295 2796
email info@americanroller.com
web address www.americanroller.com

American Roller (ARCo) manufactures rollers for a variety of industries such as printing, paper making, laminating etc. ARCo also uses thermal spray technology, particularly High Velocity Oxygen Fuel (HVOF) spraying, to deposit wear-resistant coatings on rollers to improve longevity. Both tungsten carbide and chromium carbide reinforced nickel or cobalt are offered, the metallic phase acting as a binder for the hard carbides. The surface of the roller is sandblasted before coating to remove inclusions and oxides that can result in poor adhesion of the coating. During the thermal spray process the metallic component melts and is projected with the solid carbide particles onto the roller, where the liquid metal solidifies and locks the solid particles in place. Thus, the carbide particles are embedded in the metal forming a hard composite coating.

Ametek, Inc

Specialty Metal Products
PO Box 5807, 21 Toelles Road
Wallingford CT 06492 USA

tel ++ 1 (203) 265 6731
fax ++ 1 (203) 265 6407
email smpct.custserv@ametek.com
web address www.ametekmetals.com

Ametek manufactures electrical and electromechanical devices. Ametek's Specialty Metal Products division markets metal wire, powder, and strip products, including MMC materials for thermal management applications. Under licence from AEA Technology in the UK, Ametek manufactures two grades of Al-SiC materials, HIVOL™ B and HIVOL™ C. The material is formed by pressure-assisted infiltration of silicon carbide powder preforms with aluminium. Subsequently MMC parts can be plated with nickel, gold, copper, tin or silver. The HIVOL™ materials typically contain about 68 vol. % alpha silicon carbide, but the fraction ceramic can be designed to match the thermal expansion of other materials for lightweight thermal management applications. In addition to silicon carbide-reinforced aluminium, Ametek produces molybdenum-copper and tungsten-copper materials by powder metallurgy. These materials are then rolled into sheets or plates for final machining into components such as heat sinks, thermal spreaders, chip mounts and circuit board cores. Ametek sells its thermal management products to a variety of electronics firms for applications in commercial and military sectors.

Ametek, Inc

material (company designation): HIVOL™ B

Material 2014 / SiC / 68p

matrix Al alloy

reinforcement SiC *volume percent* 68

morphology ☒ p ☐ sf ☐ cf ☐ mf ☐ w *size (μm)*

Processing

primary Infiltration

specifics pressure-assisted infiltration

secondary ☐ Casting ☐ Forging ☒ Surface Treatments ☐ Extrusion ☒ Machining ☐ Thermal Treatments

specifics

joining ☐ Brazing ☐ Mechanical ☐ Welding ☐ Other

recyclable ☐ Yes ☐ No

Properties are quoted from manufacturer as general indicators only.

Properties *isotropic* ☐ Yes ☐ No *tensile*

density (g cm^{-3}) 3.04 *yield strength (MPa)*

CTE (x10^{6} K^{-1}) 7.5 (30-150°C) *ultimate strength (MPa)* 210

thermal cond. (Wm^{-1}K^{-1}) 220 *modulus (GPa)* 228

electrical cond. (%IACS) 5.8 *elongation (%)*

Property Advantages ☒ Density ☐ Strength ☐ Wear ☐ Fatigue ☒ Thermal Expansion ☐ Weldability ☐ Stiffness ☒ Thermal Conductivity ☐ Other

Applications thermal management in electronics

Products

Sources AMETEK (2000).

Ametek, Inc

material (company designation): HIVOL™ C

Material Al-12Si / SiC / 68p

matrix Al -12 Si

reinforcement SiC *volume percent* 68

morphology ☒ p ☐ sf ☐ cf ☐ mf ☐ w *size (μm)*

Processing

primary Infiltration

specifics pressure-assisted infiltration

secondary ☐ Casting ☐ Forging ☒ Surface Treatments ☐ Extrusion ☒ Machining ☐ Thermal Treatments

specifics

joining ☐ Brazing ☐ Mechanical ☐ Welding ☐ Other

recyclable ☐ Yes ☐ No

Properties are quoted from manufacturer as general indicators only.

Properties *isotropic* ☐ Yes ☐ No *tensile*

density (g cm^{-3}) 3.03 *yield strength (MPa)*

CTE (x10^{6} K^{-1}) 6.8 (30-150°C) *ultimate strength (MPa)* 225

thermal cond. (Wm^{-1}K^{-1}) 226 *modulus (GPa)* 230

electrical cond. (%IACS) 5.4 *elongation (%)*

Property Advantages ☒ Density ☐ Strength ☐ Wear ☐ Fatigue ☒ Thermal Expansion ☐ Weldability ☐ Stiffness ☒ Thermal Conductivity ☐ Other

Applications thermal management in electronics

Products

Sources AMETEK (2000).

Applied Sciences, Inc

PO Box 579
141 West Xenia Avenue
Cedarville OH 45314-0579 USA

tel ++ 1 (937) 766 2020
fax ++ 1 (937) 766 5886
email CustomerRelations@ApSci.com
web address www.apsci.com

Applied Science, Inc (ASI) is a research and development company that specialises in advanced materials and their applications, in particular carbon-based materials and composites. ASI has developed a family of vapour-grown carbon fibres that are manufactured by its subsidiary Pyrograf® Products, Inc (PPI). These fibres have high electrical and thermal conductivity and high strength. Anticipated applications of these fibres are diverse and depend on their implementation, i.e. as individual fibres or as reinforcements in composite systems, etc.

When used as reinforcements in metals, composites with good thermal management properties are obtained. With 37 vol. % of Pyrograf-I carbon fibre reinforcement in aluminium the density is 2.44 g cm^{-3}, which is lower than most silicon carbide-reinforced aluminium composites.

Applied Sciences, Inc

material (company designation):

Material Al / carbon / 37cf

matrix unspecified grade Al

reinforcement Vapor Grown Carbon Fibre *volume percent* 37

morphology ☐ p ☐ sf ☒ cf ☐ mf ☐ w *size (μm)*

Processing

primary

specifics

secondary ☐ Casting ☐ Forging ☐ Surface Treatments ☐ Extrusion ☐ Machining ☐ Thermal Treatments

specifics

joining ☐ Brazing ☐ Mechanical ☐ Welding ☐ Other

recyclable ☐ Yes ☐ No

Properties are quoted from manufacturer as general indicators only.

Properties *isotropic* ☐ Yes ☒ No *tensile*

density ($g\ cm^{-3}$) 2.44 *yield strength (MPa)*

CTE ($x10^6\ K^{-1}$) *ultimate strength (MPa)*

thermal cond. ($Wm^{-1}K^{-1}$) 642 *modulus (GPa)*

electrical cond. (%IACS) *elongation (%)*

Property Advantages ☒ Density ☐ Strength ☐ Wear ☐ Fatigue ☐ Thermal Expansion ☐ Weldability ☐ Stiffness ☐ Thermal Conductivity ☐ Other

Applications high thermal conductivity, primarily in one direction

Products

Sources

Applied Sciences, Inc

material (company designation):

Material Cu / carbon / 50cf

matrix unspecified grade Cu

reinforcement Vapor Grown Carbon Fibre *volume percent* 50

morphology ☐ p ☐ sf ☒ cf ☐ mf ☐ w *size (μm)*

Processing

primary

specifics

secondary ☐ Casting ☐ Forging ☐ Surface Treatments ☐ Extrusion ☐ Machining ☐ Thermal Treatments

specifics

joining ☐ Brazing ☐ Mechanical ☐ Welding ☐ Other

recyclable ☐ Yes ☐ No

Properties are quoted from manufacturer as general indicators only.

Properties *isotropic* ☐ Yes ☒ No *tensile*

density ($g\ cm^{-3}$) 5.50 *yield strength (MPa)*

CTE ($x10^6\ K^{-1}$) *ultimate strength (MPa)*

thermal cond. ($Wm^{-1}K^{-1}$) 840 *modulus (GPa)*

electrical cond. (%IACS) *elongation (%)*

Property Advantages ☐ Density ☐ Strength ☐ Wear ☐ Fatigue ☐ Thermal Expansion ☐ Weldability ☐ Stiffness ☒ Thermal Conductivity ☐ Other

Applications high thermal conductivity, primarily in one direction

Products

Sources

Atlantic Research Corporation (ARC)

Atlantic IMT
220 Ballardvale Street
Wilmington MA 01887-1050 USA

tel ++ 1 (508) 658 8940
fax ++ 1 (508) 658 2306
email materials@atlanticresearchcorp.com
web address www.atlanticresearchcorp.com

Atlantic Research Corporation (ARC) is a large aerospace company that also has important activities in the automotive and advanced materials sectors. In terms of advanced materials, ARC operates in both the polymer and metal matrix composite businesses, in addition to producing silicon carbide monofilaments for titanium-based composites. Atlantic IMT, a joint venture with Bodycote IMT in Wilmington MA, manufactures titanium-matrix composites for airframe structures, turbine engines, and rocket propulsion applications. These titanium composites are reinforced by silicon carbide monofilaments and are manufactured by tape casting, wire winding and foil-fibre-foil methods, followed by consolidation using hot isostatic pressing (HIPing) at Bodycote IMT. Components in the form of panels, plates, rings and tubes have been produced. It is expected that these materials will be used in several American military applications since the manufacturing facility has obtained MIL-I-45208 qualification.

Atlantic Research Corporation (ARC)
material (company designation):

Material Ti-64 / SiC / mf

matrix Ti - 6Al - 4V

reinforcement Trimarc SiC *volume percent*

morphology ☐ p ☐ sf ☐ cf ☒ mf ☐ w *size (μm)* 127

Processing

primary Diffusion Bonding

specifics Hot Isostatic Pressing (HIPing)

secondary ☐ Casting ☐ Forging ☐ Surface Treatments ☐ Extrusion ☐ Machining ☐ Thermal Treatments

specifics

joining ☐ Brazing ☐ Mechanical ☐ Welding ☐ Other

recyclable ☐ Yes ☐ No

Properties are quoted from manufacturer as general indicators only.

Properties *isotropic* ☐ Yes ☒ No

			tensile
density (g cm^{-3})		*yield strength (MPa)*	
CTE (x10^{6} K^{-1})		*ultimate strength (MPa)*	
thermal cond. (Wm^{-1}K^{-1})		*modulus (GPa)*	
electrical cond. (%IACS)		*elongation (%)*	

Property Advantages ☐ Density ☐ Fatigue ☒ Stiffness ☒ Strength ☐ Thermal Expansion ☐ Thermal Conductivity ☐ Wear ☐ Weldability ☒ Other

Applications lightweight and high temperature applications such as turbine engines

Products

Sources

Atlantic Research Corporation (ARC)
material (company designation):

Material Ti-6242 / SiC / mf

matrix Ti - 6Al - 2Sn - 4Zr - 2Mo
reinforcement Trimarc SiC *volume percent*
morphology ☐ p ☐ sf ☐ cf ☒ mf ☐ w *size (μm)* 127

Processing
primary Diffusion Bonding
specifics Hot Isostatic Pressing (HIPing)

secondary ☐ Casting ☐ Forging ☐ Surface Treatments ☐ Extrusion ☐ Machining ☐ Thermal Treatments
specifics

joining ☐ Brazing ☐ Mechanical ☐ Welding ☐ Other
recyclable ☐ Yes ☐ No

Properties are quoted from manufacturer as general indicators only.

Properties *isotropic* ☐ Yes ☒ No *tensile*
density ($g\ cm^{-3}$) *yield strength (MPa)*
CTE ($x10^{6}\ K^{-1}$) *ultimate strength (MPa)*
thermal cond. ($Wm^{-1}K^{-1}$) *modulus (GPa)*
electrical cond. (%IACS) *elongation (%)*

Property Advantages ☐ Density ☒ Strength ☐ Wear ☐ Fatigue ☐ Thermal Expansion ☐ Weldability ☒ Stiffness ☐ Thermal Conductivity ☒ Other

Applications lightweight and high temperature applications such as turbine engines

Products

Sources

Atlantic Research Corporation (ARC)
material (company designation):

Material Ti55Al22Nb23 / SiC / mf

matrix Ti - 22Al - 23Nb (at%)

reinforcement Trimarc SiC *volume percent*

morphology ☐ p ☐ sf ☐ cf ☒ mf ☐ w *size (μm)* 127

Processing

primary Diffusion Bonding

specifics Hot Isostatic Pressing (HIPing)

secondary ☐ Casting ☐ Forging ☐ Surface Treatments ☐ Extrusion ☐ Machining ☐ Thermal Treatments

specifics

joining ☐ Brazing ☐ Mechanical ☐ Welding ☐ Other

recyclable ☐ Yes ☐ No

Properties are quoted from manufacturer as general indicators only.

Properties *isotropic* ☐ Yes ☒ No *tensile*

density ($g\ cm^{-3}$) *yield strength (MPa)*

CTE ($x10^6\ K^{-1}$) *ultimate strength (MPa)*

thermal cond. ($Wm^{-1}K^{-1}$) *modulus (GPa)*

electrical cond. (%IACS) *elongation (%)*

Property Advantages ☐ Density ☒ Strength ☐ Wear ☐ Fatigue ☐ Thermal Expansion ☐ Weldability ☒ Stiffness ☐ Thermal Conductivity ☒ Other

Applications lightweight and high temperature applications such as turbine engines

Products

Sources

Bell Helicopter Textron, Inc

PO Box 482

Fort Worth TX 76101 USA

tel ++ 1 (817) 280 2011
fax ++ 1 (817) 280 2321
email
web address www.bellhelicopter.textron.com

Bell Helicopter Textron is a major producer of commercial and military helicopters. In collaboration with Boeing, Bell manufactures the Bell Boeing V-22 Osprey tiltrotor helicopter for the US armed services. The project has delivered at least 4 helicopters and they continue to conduct operational tests and evaluations before the helicopter will enter full-scale production.

Initially, whisker-reinforced aluminium was employed for a hydraulic manifold on the V-22. This manifold redistributes high-pressure hydraulic fluid (up to 70 MPa) to flight control surfaces. Aluminium-matrix composite was chosen because aluminium alloy does not exhibit sufficient burst strength. The manifold is manufactured by machining from a solid piece, a costly procedure. Hence, to reduce costs, Triton Systems prepared a near net-shape component by infiltration (A206/SiC/40p).

Bodycote IMT, Inc

155 River Street

Andover MA 01810 USA

tel ++ 1 (978) 470 1620
fax ++ 1 (978) 475 2951
email sales@bodycote-imt.com
web address www.bodycote.com

Bodycote IMT is a division of Bodycote International plc, a metallurgical service company with a global reach. Bodycote uses hot isostatic pressing (HIPing) to manufacture MMC. A variety of material systems can be employed with this powder metallurgical method, including tool steels reinforced with carbides, nitrides, oxides, borides and silicides, or light alloys reinforced with ceramic particles. HIPing has also been investigated for the manufacture of stainless steel wire reinforced aluminium alloys, also called metal-reinforced metal matrix composites (MMMC). The powder metallurgy approach reduces interfacial reactions compared to liquid processing routes such as infiltration. Components can also be clad with MMC materials to improve wear resistance such as cladding an aluminium structure with an aluminium-silicon carbide mixture.

in Europe:

Bodycote HIP, Ltd
Sheffield Road
Sheepbridge
Chesterfield, S41 9ED
UK

tel ++44 (1246) 260 456
fax ++44 (1246) 260 889

(updated Feb. 2003)

Boeing

Military Aircraft and Missile Systems
PO Box 516
St. Louis MO 63166 USA

tel ++ 1 (314) 232 0232
fax ++
email
web address www.boeing.com

Boeing has a diverse range of aerospace products, and MMC components are filling important needs in high performance applications.

The F/A-18-E/F Super Hornet, an aircraft carrier based multi-role fighter and attack aircraft, is one of the US Navy's newest aircraft. Hydraulic components in the landing gear of this aircraft are produced from 2009/SiC/15.5p material from DWA Aluminum Composites. The F/A-18 E/F was produced by McDonnell Douglas before it was acquired by Boeing.

Boeing, in a partnership with Bell Helicopter Textron, is responsible for the fuselage, digital avionics, and fly-by-wire flight-control systems on the V-22 Osprey tiltrotor aircraft. A MMC (Al/SiC/40p) hydraulic manifold is employed on V-22; the component is produced by infiltration by Triton Systems.

MMC materials are employed on the AC-130U Gunship as ammunition racks; weight reduction led to the choice of particle-reinforced aluminium racks. Boeing is a copartner with Lockheed Martin on the AC-130U.

These MMC materials also meet the needs of commercial aircraft. Pratt and Whitney's 4000 series engines for Boeing's 777 aircraft are equipped with Fan Exit Guide Vanes (FEGVs) produced from 6092/SiC/17.5 supplied by DWA.

Thermal management is a growing need for high-density electronics, particularly onboard military aircraft and spacecraft. A number of aircraft in the Boeing family reportedly utilise MMC materials from dmc2 Electronic Components (DEC) for thermal management, such as for the heads-up-display panel, fly-by-wire controls and the power generation unit in the F-22 Raptor.

Brush Wellman Inc

14710 West Portage River South Road

Elmore OH 43416-9502 USA

tel ++ 1 (419) 862 4171
fax ++ 1 (419) 862 4174
email beproducts@brushwellman.com
web address www.brushwellman.com

Brush Wellman is a major supplier of beryllium-based materials. Two distinct types of MMC are produced by Brush Wellman: beryllium oxide-reinforced beryllium (Be-BeO) and beryllium-reinforced aluminium (AlBeMet®). Both composite systems are manufactured by powder metallurgical routes using hot isostatic pressing (HIPing) or cold isostatic pressing (CIPing).

The Be-BeO materials, called E-Materials, are offered in three standard grades based on concentration of the BeO platelets, although the concentration can be tailored for specific requirements.

AlBeMet® has high concentrations of beryllium in an aluminium matrix, and is produced by consolidation of gas-atomised powders. Conventional metals processing techniques are employed after consolidation, such as extrusion to produce profiles or rolling to produce sheet. AlBeMet® 162 is a "reincarnation of Lockalloy" developed in the 1960s by Lockheed for a folding ventral fin on the SR-71 Blackbird (YF-12a). Containing about 70 vol. % Be, AlBeMet® 162 has attractive properties for thermal management solutions and for lightweight structural components. The composition of AlBeMet® can be modified for specific needs and an investment casting alloy, AlBeCast®, has also been developed.

Honeycomb panels on the Orbcomm satellites are constructed from AlBeMet® sheet; the panels are fastened by three brackets produced from AlBeMet® extrusions. In addition, AlBeMet® is used for the support structure for the antenna boom segments in the form of gusset panels and rectangular tubes.

Thermal management in the avionics sector is an important thrust for Brush Wellman and both E-Materials and AlBeMet® are appropriate in these applications. There has been a focussed effort to implement AlBeMet® for the metal cores that support and cool the electronics of the central computer on the US Air Force F-22 Raptor aircraft. Over a thousand such cores have been manufactured and fully qualified.

Brush Wellman Inc

material (company designation): E20 E-material

Material Be / BeO / 20p

matrix Be

reinforcement BeO *volume percent* 20

morphology ☒ p ☐ sf ☐ cf ☐ mf ☐ w *size (μm)*

Processing

primary Powder Metallurgy

specifics Hot Isostatic Pressing (HIPing); Cold Isostatic

secondary ☐ Casting ☐ Forging ☐ Surface Treatments ☐ Extrusion ☒ Machining ☒ Thermal Treatments

specifics

joining ☒ Brazing ☐ Mechanical ☐ Welding ☐ Other

recyclable ☒ Yes ☐ No

Properties are quoted from manufacturer as general indicators only.

Properties *isotropic* ☒ Yes ☐ No *tensile*

density ($g\ cm^{-3}$) 2.06 *yield strength (MPa)*

CTE ($x10^6\ K^{-1}$) 8.7 (25-50°C ave) *ultimate strength (MPa)* 400

thermal cond. ($Wm^{-1}K^{-1}$) 210 *modulus (GPa)* 303

electrical cond. (%IACS) *elongation (%)*

Property Advantages ☒ Density ☐ Strength ☐ Wear ☒ Fatigue ☒ Thermal Expansion ☐ Weldability ☒ Stiffness ☒ Thermal Conductivity ☒ Other

Applications thermal managament in electronics

Products

Sources Brush Wellman Inc (2000).
Brush Wellman Inc.

Brush Wellman Inc

material (company designation): E40 E-material

Material Be / BeO / 40p

matrix Be

reinforcement BeO *volume percent* 40

morphology ☒ p ☐ sf ☐ cf ☐ mf ☐ w *size (μm)*

Processing

primary Powder Metallurgy

specifics Hot Isostatic Pressing (HIPing); Cold Isostatic

secondary ☐ Casting ☐ Forging ☐ Surface Treatments ☐ Extrusion ☒ Machining ☐ Thermal Treatments

specifics

joining ☒ Brazing ☐ Mechanical ☐ Welding ☐ Other

recyclable ☒ Yes ☐ No

Properties are quoted from manufacturer as general indicators only.

Properties *isotropic* ☒ Yes ☐ No *tensile*

density (g cm⁻³) 2.3 *yield strength (MPa)*

CTE (x10⁶ K⁻¹) 7.5 (25-50°C ave) *ultimate strength (MPa)* 290

thermal cond. (Wm⁻¹K⁻¹) 220 *modulus (GPa)* 317

electrical cond. (%IACS) *elongation (%)*

Property Advantages ☒ Density ☐ Strength ☐ Wear ☒ Fatigue ☒ Thermal Expansion ☐ Weldability ☒ Stiffness ☒ Thermal Conductivity ☒ Other

Applications thermal managament in electronics

Products

Sources Brush Wellman Inc (2000).
Brush Wellman Inc.

Brush Wellman Inc

material (company designation): E60 E-material

Material Be / BeO / 60p

matrix Be

reinforcement BeO *volume percent* 60

morphology ☒ p ☐ sf ☐ cf ☐ mf ☐ w *size (μm)*

Processing

primary Powder Metallurgy

specifics Hot Isostatic Pressing (HIPing); Cold Isostatic

secondary ☐ Casting ☐ Forging ☐ Surface Treatments ☐ Extrusion ☒ Machining ☒ Thermal Treatments

specifics

joining ☒ Brazing ☐ Mechanical ☐ Welding ☐ Other

recyclable ☒ Yes ☐ No

Properties are quoted from manufacturer as general indicators only.

Properties *isotropic* ☒ Yes ☐ No *tensile*

density (g cm^{-3}) 2.52 *yield strength (MPa)*

CTE (x10^{6} K^{-1}) 6.1 (25-50°C ave) *ultimate strength (MPa)* 270

thermal cond. (Wm^{-1}K^{-1}) 230 *modulus (GPa)* 330

electrical cond. (%IACS) *elongation (%)*

Property Advantages ☒ Density ☐ Strength ☐ Wear ☒ Fatigue ☒ Thermal Expansion ☐ Weldability ☒ Stiffness ☒ Thermal Conductivity ☒ Other

Applications thermal managament in electronics

Products

Sources Brush Wellman Inc (2000).

Brush Wellman Inc

material (company designation): AM162 (AlBeMet®)

Material Al / Be / 70p / extruded / annealed

matrix unspecified grade Al

reinforcement 62 wt% Be *volume percent* 70

morphology ☒ p ☐ sf ☐ cf ☐ mf ☐ w *size (μm)*

Processing

primary Powder Metallurgy

specifics Hot Isostatic Pressing (HIPing); Cold Isostatic

secondary ☒ Casting ☐ Forging ☐ Surface Treatments ☒ Extrusion ☒ Machining ☐ Thermal Treatments

specifics

joining ☒ Brazing ☒ Mechanical ☒ Welding ☐ Other

recyclable ☒ Yes ☐ No

Properties are quoted from manufacturer as general indicators only.

Properties *isotropic* ☐ Yes ☐ No *tensile*

density ($g\ cm^{-3}$) 2.1 *yield strength (MPa)* 276

CTE ($x10^6\ K^{-1}$) 13.9 (25°C) *ultimate strength (MPa)* 380

thermal cond. ($Wm^{-1}K^{-1}$) 210 *modulus (GPa)* 195

electrical cond. (%IACS) *elongation (%)* 7

Property Advantages ☒ Density ☐ Strength ☐ Wear ☒ Fatigue ☒ Thermal Expansion ☐ Weldability ☒ Stiffness ☒ Thermal Conductivity ☒ Other

Applications heatsinks; structural components such as gimbals; similar to Lockalloy developed by Lockheed for the folding ventral fin of the SR-71 Blackbird (Lockheed YF-12a); brake calliper for Ferrari Formula 1; avionics suite for Lockheed F-22 Raptor aircraft

Products communications satellite structural frame

Sources Brush Wellman Inc (2000).
Brush Wellman Inc.

Brush Wellman Inc

material (company designation): AlBeCast®

Material Al / Be / 38p

matrix unspecified grade Al

reinforcement 30 wt% Be *volume percent* 38

morphology ☒ p ☐ sf ☐ cf ☐ mf ☐ w *size (μm)*

Processing

primary

specifics Hot Isostatic Pressing (HIPing); Cold Isostatic

secondary ☐ Casting ☐ Forging ☐ Surface Treatments ☐ Extrusion ☐ Machining ☐ Thermal Treatments

specifics

joining ☐ Brazing ☐ Mechanical ☐ Welding ☐ Other

recyclable ☐ Yes ☐ No

Properties are quoted from manufacturer as general indicators only.

Properties *isotropic* ☐ Yes ☐ No

density $(g\ cm^{-3})$	2.17		*tensile*
		yield strength (MPa)	158
CTE $(x10^{6}\ K^{-1})$	14.6 (20-100°C)	*ultimate strength (MPa)*	207
thermal cond. $(Wm^{-1}K^{-1})$	110	*modulus (GPa)*	193
electrical cond. (%IACS)		*elongation (%)*	4

Property Advantages ☒ Density ☐ Fatigue ☒ Stiffness ☐ Strength ☒ Thermal Expansion ☒ Thermal Conductivity ☐ Wear ☐ Weldability ☐ Other

Applications casting version of AlBeMet®

Products

Sources

Capral Aluminium Ltd

PO Box 12

Granville NSW 2142 Australia

tel ++ 61 (2) 9682 0711
fax ++ 61 (2) 9682 0777
email contact@capral.com.au
web address www.capral-aluminium.com.au

Capral Aluminium produces sheet and extruded products for the automotive industry. Capral extruded Duralcan (from Alcan) producing tubing for the propshaft employed by General Motors for the Chevrolet Corvette. "Duralcan" tubing is 40% stiffer than aluminium concomitant with a reduction in thermal expansion of 18% and improved wear resistance of 89%. Although the property advantages are clear, the economics of manufacturing the Corvette propshaft from MMC proved unfavourable in the long term; hence, it is currently produced from 6061 aluminium. Nevertheless, this represents a significant step forward in the implementation of MMC into mass production automobile, and even though production of MMC propshafts for the Corvette has ceased, Capral retains the technology, capability and experience to manufacture MMC tubing in the future.

Ceramics Process Systems (CPS)

111 South Worcester Street

Chartley MA 02712-0338 USA

tel ++ 1 (508) 222 0614
fax ++ 1 (508) 222 0220
email kvalen@alsic.com
web address www.alsic.com

Ceramic Process Systems (CPS) manufactures thermal management and electronic packaging components produced from the Al-SiC family of materials. Silicon carbide preforms are prepared using a low pressure injection molding process called QuickSet™ and they are subsequently infiltrated with molten aluminum by the Quickcast™ process to produce near-net shape components. In addition, CPS has developed Concurrent Integration™, whereby functional components such as feedthroughs, sealrings, and substrates can be hermetically integrated during fabrication of the MMC component. This eliminates the need for subsequent brazing or soldering.

CPS manufactures baseplates and substrates that require sufficient thermal conductivity to dissipate the heat from high-density electronic circuits, in addition to a coefficient of thermal expansion (CTE) to match the module design. Convex bow shapes can be achieved by CPS's near-net shape processing, thus enhancing module-baseplate interfacing for heat dissipation while maintaining flatness for attaching the dielectric substrate. The high stiffness of Al-SiC materials allows cooling devices to incorporate pin fins to enhance the heat dissipation capability of these devices. CPS produces components from three grades of Al-SiC material, each tailored for specific applications. AlSiC-9 is suitable for ceramic Flip Chip lid assemblies, direct attachment to GaAs modules, IGBT (Insulated Gate Bipolar Transistor) thermal base plates, and thermal substrates. AlSiC-10 and AlSiC-12 have lower concentrations of SiC, making them suitable as plastic Flip Chip lid assemblies. CPS also supplies Al-SiC components for high-density microprocessor assemblies, wireless base stations, motor controllers, and satellite communications.

Ceramics Process Systems (CPS)

material (company designation): AlSiC-8

Material A356.2 / SiC / 67p

matrix Al alloy

reinforcement SiC *volume percent* 67

morphology ☒ p ☐ sf ☐ cf ☐ mf ☐ w *size (μm)*

Processing

primary Infiltration

specifics QuickCast™ infiltration process, Concurrent

secondary ☐ Casting ☐ Forging ☒ Surface Treatments ☐ Extrusion ☐ Machining ☐ Thermal Treatments

specifics plating, anodizing

joining ☒ Brazing ☐ Mechanical ☐ Welding ☐ Other

recyclable ☐ Yes ☐ No

Properties are quoted from manufacturer as general indicators only.

Properties *isotropic* ☒ Yes ☐ No *tensile*

density (g cm^{-3}) 2.99 *yield strength (MPa)*

CTE (x10^{6} K^{-1}) 6.88 (35-100°C) *ultimate strength (MPa)*

thermal cond. (Wm^{-1}K^{-1}) 180 *modulus (GPa)* 217.5

electrical cond. (%IACS) *elongation (%)*

Property Advantages ☒ Density ☐ Strength ☐ Wear ☐ Fatigue ☒ Thermal Expansion ☐ Weldability ☐ Stiffness ☒ Thermal Conductivity ☐ Other

Applications power IC housings, microwave housings, finned heatsinks, carrier plates, microprocessor lids

Products

Sources Ceramics Process Systems (1999).

Ceramics Process Systems (CPS)
material (company designation): AlSiC-9

Material	A356.2 / SiC / 63p		
matrix	Al alloy		
reinforcement	SiC	*volume percent*	63
morphology	☒ p ☐ sf ☐ cf ☐ mf ☐ w	*size (μm)*	

Processing	
primary	Infiltration
specifics	QuickCast™ infiltration process, Concurrent
secondary	☐ Casting ☐ Forging ☒ Surface Treatments ☐ Extrusion ☐ Machining ☐ Thermal Treatments
specifics	plating, anodizing
joining	☒ Brazing ☐ Mechanical ☐ Welding ☐ Other
recyclable	☐ Yes ☐ No

Properties are quoted from manufacturer as general indicators only.

Properties	*isotropic* ☒ Yes ☐ No		*tensile*
density ($g\ cm^{-3}$)	3.01	*yield strength (MPa)*	
CTE ($x10^{6}\ K^{-1}$)	7.96	*ultimate strength (MPa)*	
thermal cond. ($Wm^{-1}K^{-1}$)	170 minimum	*modulus (GPa)*	192
electrical cond. (%IACS)		*elongation (%)*	

Property Advantages	☒ Density	☐ Strength	☐ Wear
	☐ Fatigue	☒ Thermal Expansion	☐ Weldability
	☐ Stiffness	☒ Thermal Conductivity	☐ Other

Applications	power IC housings, microwave housings, finned heatsinks, carrier plates, microprocessor lids
Products	
Sources	Ceramics Process Systems (1999).

Ceramics Process Systems (CPS)

material (company designation): AlSiC-10

Material A356.2 / SiC / 54p

matrix Al alloy

reinforcement SiC *volume percent* 54

morphology ☒ p ☐ sf ☐ cf ☐ mf ☐ w *size (μm)*

Processing

primary Infiltration

specifics QuickCast™ infiltration process, Concurrent

secondary ☐ Casting ☐ Forging ☒ Surface Treatments ☐ Extrusion ☐ Machining ☐ Thermal Treatments

specifics plating, anodizing

joining ☒ Brazing ☐ Mechanical ☐ Welding ☐ Other

recyclable ☐ Yes ☐ No

Properties are quoted from manufacturer as general indicators only.

Properties *isotropic* ☒ Yes ☐ No *tensile*

density ($g\ cm^{-3}$) 2.96 *yield strength (MPa)*

CTE ($x10^6\ K^{-1}$) 9.75 (30-100°C) *ultimate strength (MPa)*

thermal cond. ($Wm^{-1}K^{-1}$) 170 minimum *modulus (GPa)* 167

electrical cond. (%IACS) *elongation (%)*

Property Advantages ☒ Density ☐ Strength ☐ Wear ☐ Fatigue ☒ Thermal Expansion ☐ Weldability ☐ Stiffness ☒ Thermal Conductivity ☐ Other

Applications power IC housings, microwave housings, finned heatsinks, carrier plates, microprocessor lids

Products

Sources Ceramics Process Systems (1999).

Chesapeake Composites Corporation
239 Old Churchman's Road
New Castle DE 19720 USA
tel ++ 1 (302) 324 9110
fax ++ 1 (302) 324 0886
email info@chesapeakecomposites.com
web address www.chesapeakecomposites.com

Chesapeake Composites manufactures DSC Aluminum, an aluminium reinforced by sub-micrometer-sized alumina particles. This material combines the improved stiffness of conventional composites with the advantages of dispersion-strengthened materials. High concentrations of alumina (around 35 vol. %) give the material its composite character, while the very fine particles, or dispersoids (about 0.3 μm in diameter) provide DSC Aluminum with good high temperature strength as typically associated with dispersion-strengthened metals. Toughness and ductility are provided primarily by the pure aluminium matrix, and perhaps less importantly by the small size of the reinforcement. DSC Aluminium has higher specific stiffness compared with other metals, and superior creep resistance and high-temperature strength compared to both conventional and particle-reinforced aluminium alloys.

DSC Aluminum is fabricated by a patented process (US Patents 5,702,542 and 5,511,603): pressure infiltration of a preform of highly uniform, low-cost, commodity grade, calcinated alumina dispersoids. This procedure produces a billet that is ready for secondary processing operations such as extrusion or semi-solid forming. By virtue of the soft aluminium matrix and the small size of reinforcing particles, this material can be machined using conventional-point carbide tools and procedures, a distinct benefit compared to most other ceramic reinforced metals.

Chesapeake Composites Corporation
material (company designation):

Material 1090 / Al2O3 / 30p / extrusion / F

matrix pure Al
reinforcement Al2O3 *volume percent* 30
morphology ☒ p ☐ sf ☐ cf ☐ mf ☐ w *size (μm)*

Processing
primary Infiltration
specifics

secondary ☐ Casting ☒ Forging ☐ Surface Treatments
☒ Extrusion ☒ Machining ☐ Thermal Treatments

specifics

joining ☐ Brazing ☐ Mechanical ☐ Welding ☐ Other
recyclable ☐ Yes ☐ No

Properties are quoted from manufacturer as general indicators only.

Properties *isotropic* ☐ Yes ☐ No *tensile*

density (g cm⁻³)	3.08	*yield strength (MPa)*	292
CTE (x10⁶ K⁻¹)	16.5	*ultimate strength (MPa)*	432
thermal cond. (Wm⁻¹K⁻¹)	140	*modulus (GPa)*	115
electrical cond. (%IACS)		*elongation (%)*	8

Property Advantages ☒ Density ☒ Strength ☐ Wear
☐ Fatigue ☐ Thermal Expansion ☐ Weldability
☒ Stiffness ☐ Thermal Conductivity ☒ Other

Applications machinable aluminium-based composite with good creep resistance

Products

Sources

Chesapeake Composites Corporation
material (company designation):

Material 1090 / Al2O3 / 40p / extrusion / F

matrix pure Al

reinforcement Al2O3 *volume percent* 40

morphology ☒ p ☐ sf ☐ cf ☐ mf ☐ w *size (μm)*

Processing

primary Infiltration

specifics

secondary ☐ Casting ☒ Forging ☐ Surface Treatments ☒ Extrusion ☒ Machining ☐ Thermal Treatments

specifics

joining ☐ Brazing ☐ Mechanical ☐ Welding ☐ Other

recyclable ☐ Yes ☐ No

Properties are quoted from manufacturer as general indicators only.

Properties *isotropic* ☐ Yes ☐ No

			tensile
density (*g cm*$^{-3}$)	3.21	*yield strength (MPa)*	427
CTE ($x10^6$ K^{-1})	15.6	*ultimate strength (MPa)*	530
thermal cond. ($Wm^{-1}K^{-1}$)	111	*modulus (GPa)*	132
electrical cond. (%IACS)		*elongation (%)*	4.5

Property Advantages ☒ Density ☒ Strength ☐ Wear ☐ Fatigue ☐ Thermal Expansion ☐ Weldability ☒ Stiffness ☐ Thermal Conductivity ☒ Other

Applications machinable aluminium-based composite with good creep resistance

Products

Sources

Composite Metal Technology Ltd (CMT)

Griffith Building (A7)
Cody Technology Park
Farnborough, Hampshire GU14 0LX UK

tel ++ 44 (1252) 397 447
fax ++ 44 (1252) 397 315
email cmt@taz.qinetiq.com
web address www.cmt-ltd.com

Composite Metal Technology (CMT) is the sole licencee of the patented Liquid Pressure Forming (LPF) process, developed within governmental agencies in the UK. The LPF technology is being commercialised by the manufacture of LPF fast-cycle machines (MCM series 2). MIR SpA, a large Italian machine manufacturer, has assisted CMT in development of this technology and has a 15-year agreement to manufacture MCM machines.

LPF fast-cycle machines are reportedly capable of infiltrating any type of porous medium with molten aluminium, magnesium or zinc. Preheated preforms are placed into the heated die within the LPF fast cycle machine, which is closed and locked using a mechanical toggle system based on a MIR design. Both the die and the pressure vessel containing the furnace are then evacuated in about 10 seconds. Nitrogen gas introduced into the pressure vessel forces the molten alloy up a riser tube from the crucible into the die and the preform. Infiltration is completed with the assistance of two hydraulic pistons that pressurise the molten metal to high pressures. The total cycle time is under 2 minutes but can be lowered with multi-cavity dies.

This process provides some important advantages for processing MMC: the evacuation step ensures minimal porosity from gas entrapped within the composite, the two-stage pressurisation/infiltration step reduces preform damage, the pistons assist control of grain size as well as transmitting high-pressure for infiltration, it is not necessary to significantly superheat the melt, and the volume of melt is controlled reducing waste. The addition of standard equipment suppliers and technology expertise for infiltration will likely contribute to future commercialisation of MMC products. Although presently CMT does not produce MMC commercially, several components for transport and ordinance markets are being pursued.

Contraves Space AG

Schaffhauserstrasse 580

Zürich 8052 Switzerland

tel ++ 41 (1) 306 22 11
fax ++ 41 (1) 306 29 10
email helga.schib@unaxis.com
web address www.contravesspace.com

Contraves Space provides payload fairings for spacecraft launch vehicles, spacecraft structures, and subsystems for space platforms, in addition to other needs of space-based programs. To meet stringent payload specifications, Contraves Space uses AlBeMet® material (beryllium-reinforced aluminium composites) from Brush Wellman, and other MMC systems are being assessed.

Controltech

Controltech Bicycle Components
1840 Terminal Drive
Richland WA 99352 USA

tel ++ 1 (509) 946 7461 // 1 (877) 929 0800
fax ++ 1 (509) 946 7663
email information@controltechbikes.com
web address www.controltechbikes.com

Controltech uses innovative design and materials to produce components for a number of industries, including high-performance cycling. The superior strength and the excellent abrasion resistance are the key benefits of using MMC for bicycle rims. These rims are expected to have a lifetime of four times that of conventional aluminium rims. These lightweight (388g) and relatively inexpensive (90 USD) rims have reportedly generated sales of over 1.4 million USD. The technology for producing MMC rims was developed by Spencer Technology and it has been targeted for wheelchairs as well.

In the future, Controltech anticipates manufacturing MMC brake discs.

Creuzet Aeronautique

Route de Beyssac

Marmande 47200 France

tel ++ 33 (553) 20 45 00
fax ++ 33 (553) 20 45 20
email creuzet@creuzet.fr
web address www.creuzet.fr

Creuzet specialises in extrusion of light alloys for a range of aeronautical applications and has extensive experience extruding particle-reinforced aluminium. Creuzet can supply extrusions of 2000, 6000 and 8000 series aluminium alloys reinforced with varying volume fractions of silicon carbide. Creuzet has not found a market for MMC extrusions; however, the capacity and capability remain.

Creuzet Aeronautique

material (company designation):

Material 2124 / SiC / 17p / T4

matrix Al alloy

reinforcement SiC *volume percent* 17

morphology ☒ p ☐ sf ☐ cf ☐ mf ☐ w *size (μm)*

Processing

primary

specifics

secondary ☐ Casting ☐ Forging ☐ Surface Treatments ☒ Extrusion ☒ Machining ☒ Thermal Treatments

specifics

joining ☐ Brazing ☐ Mechanical ☐ Welding ☐ Other

recyclable ☐ Yes ☐ No

Properties are quoted from manufacturer as general indicators only.

Properties *isotropic* ☐ Yes ☐ No *tensile*

density (g cm^{-3}) 2.85 *yield strength (MPa)* 400

CTE (x10^6 K^{-1}) *ultimate strength (MPa)* 610

thermal cond. (Wm^{-1}K^{-1}) *modulus (GPa)* 100

electrical cond. (%IACS) *elongation (%)* 7

Property Advantages ☒ Density ☒ Strength ☐ Wear ☐ Fatigue ☐ Thermal Expansion ☐ Weldability ☒ Stiffness ☐ Thermal Conductivity ☐ Other

Applications extrusions with high specific stiffness

Products

Sources

Creuzet Aeronautique

material (company designation):

Material 2124 / SiC / 17p / T351

matrix Al alloy

reinforcement SiC *volume percent* 17

morphology ☒ p ☐ sf ☐ cf ☐ mf ☐ w *size (μm)*

Processing

primary

specifics

secondary ☐ Casting ☐ Forging ☐ Surface Treatments ☒ Extrusion ☒ Machining ☒ Thermal Treatments

specifics

joining ☐ Brazing ☐ Mechanical ☐ Welding ☐ Other

recyclable ☐ Yes ☐ No

Properties are quoted from manufacturer as general indicators only.

Properties *isotropic* ☐ Yes ☐ No

			tensile
density ($g\ cm^{-3}$)	2.85	*yield strength (MPa)*	500
CTE ($x10^6\ K^{-1}$)		*ultimate strength (MPa)*	610
thermal cond. ($Wm^{-1}K^{-1}$)		*modulus (GPa)*	100
electrical cond. (%IACS)		*elongation (%)*	6

Property Advantages ☒ Density ☒ Strength ☐ Wear ☐ Fatigue ☐ Thermal Expansion ☐ Weldability ☒ Stiffness ☐ Thermal Conductivity ☐ Other

Applications extrusions with high specific stiffness

Products

Sources

Creuzet Aeronautique

material (company designation):

Material 2124 / SiC / 25p / T4

matrix Al alloy

reinforcement SiC *volume percent* 25

morphology ☒ p ☐ sf ☐ cf ☐ mf ☐ w *size (μm)*

Processing

primary

specifics

secondary ☐ Casting ☐ Forging ☐ Surface Treatments ☒ Extrusion ☒ Machining ☒ Thermal Treatments

specifics

joining ☐ Brazing ☐ Mechanical ☐ Welding ☐ Other

recyclable ☐ Yes ☐ No

Properties are quoted from manufacturer as general indicators only.

Properties *isotropic* ☐ Yes ☐ No

density (g cm^{-3})	2.88	*tensile yield strength (MPa)*	500
CTE (x10^{6} K^{-1})		*ultimate strength (MPa)*	700
thermal cond. (Wm^{-1}K^{-1})		*modulus (GPa)*	115
electrical cond. (%IACS)		*elongation (%)*	4

Property Advantages ☒ Density ☒ Strength ☐ Wear ☐ Fatigue ☐ Thermal Expansion ☐ Weldability ☒ Stiffness ☐ Thermal Conductivity ☐ Other

Applications extrusions with high specific stiffness

Products

Sources

Creuzet Aeronautique

material (company designation):

Material 6013 / SiC / 20p / T6

matrix Al alloy

reinforcement SiC *volume percent* 20

morphology ☒ p ☐ sf ☐ cf ☐ mf ☐ w *size (μm)*

Processing

primary

specifics

secondary ☐ Casting ☐ Forging ☐ Surface Treatments ☒ Extrusion ☒ Machining ☒ Thermal Treatments

specifics

joining ☐ Brazing ☐ Mechanical ☐ Welding ☐ Other

recyclable ☐ Yes ☐ No

Properties are quoted from manufacturer as general indicators only.

Properties *isotropic* ☐ Yes ☐ No

density (g cm^{-3})	2.82		*tensile*
CTE (x10^{6} K^{-1})		*yield strength (MPa)*	450
thermal cond. (Wm^{-1}K^{-1})		*ultimate strength (MPa)*	520
electrical cond. (%IACS)		*modulus (GPa)*	104
		elongation (%)	5

Property Advantages ☒ Density ☐ Fatigue ☒ Stiffness ☒ Strength ☐ Thermal Expansion ☐ Thermal Conductivity ☐ Wear ☐ Weldability ☐ Other

Applications extrusions with high specific stiffness

Products

Sources

Creuzet Aeronautique

material (company designation):

Material 6113 / SiC / 25p / T6

matrix Al alloy

reinforcement SiC *volume percent* 25

morphology ☒ p ☐ sf ☐ cf ☐ mf ☐ w *size (μm)*

Processing

primary

specifics

secondary ☐ Casting ☐ Forging ☐ Surface Treatments ☒ Extrusion ☒ Machining ☒ Thermal Treatments

specifics

joining ☐ Brazing ☐ Mechanical ☐ Welding ☐ Other

recyclable ☐ Yes ☐ No

Properties are quoted from manufacturer as general indicators only.

Properties *isotropic* ☐ Yes ☐ No *tensile*

density (g cm^{-3}) 2.84 *yield strength (MPa)* 490

CTE (x10^{6} K^{-1}) *ultimate strength (MPa)* 540

thermal cond. (Wm^{-1}K^{-1}) *modulus (GPa)* 121

electrical cond. (%IACS) *elongation (%)* 2

Property Advantages ☒ Density ☒ Strength ☐ Wear ☐ Fatigue ☐ Thermal Expansion ☐ Weldability ☒ Stiffness ☐ Thermal Conductivity ☐ Other

Applications extrusions with high specific stiffness

Products

Sources

Creuzet Aeronautique

material (company designation):

Material 6092 / SiC / 25p / T651

matrix Al alloy

reinforcement SiC *volume percent* 25

morphology ☒ p ☐ sf ☐ cf ☐ mf ☐ w *size (μm)*

Processing

primary

specifics

secondary ☐ Casting ☐ Forging ☐ Surface Treatments
☒ Extrusion ☒ Machining ☒ Thermal Treatments

specifics

joining ☐ Brazing ☐ Mechanical ☐ Welding ☐ Other

recyclable ☐ Yes ☐ No

Properties are quoted from manufacturer as general indicators only.

Properties *isotropic* ☐ Yes ☐ No *tensile*

density (g cm^{-3}) 2.84 *yield strength (MPa)* 480

CTE (x10^{6} K^{-1}) *ultimate strength (MPa)* 520

thermal cond. (Wm^{-1}K^{-1}) *modulus (GPa)*

electrical cond. (%IACS) *elongation (%)* 3

Property Advantages ☒ Density ☒ Strength ☐ Wear
☐ Fatigue ☐ Thermal Expansion ☐ Weldability
☒ Stiffness ☐ Thermal Conductivity ☐ Other

Applications extrusions with high specific stiffness

Products

Sources

Creuzet Aeronautique

material (company designation):

Material 8090 / SiC / 17p / T6

matrix Al-Li

reinforcement SiC *volume percent* 17

morphology ☒ p ☐ sf ☐ cf ☐ mf ☐ w *size (μm)*

Processing

primary

specifics

secondary ☐ Casting ☐ Forging ☐ Surface Treatments ☒ Extrusion ☒ Machining ☒ Thermal Treatments

specifics

joining ☐ Brazing ☐ Mechanical ☐ Welding ☐ Other

recyclable ☐ Yes ☐ No

Properties are quoted from manufacturer as general indicators only.

Properties *isotropic* ☐ Yes ☐ No

density $(g\ cm^{-3})$	2.66		*tensile*
CTE $(x10^6\ K^{-1})$		*yield strength (MPa)*	450
thermal cond. $(Wm^{-1}K^{-1})$		*ultimate strength (MPa)*	540
electrical cond. (%IACS)		*modulus (GPa)*	103
		elongation (%)	4

Property Advantages ☒ Density ☐ Fatigue ☒ Stiffness ☒ Strength ☐ Thermal Expansion ☐ Thermal Conductivity ☐ Wear ☐ Weldability ☐ Other

Applications extrusions with high specific stiffness

Products

Sources

CSM Industries

21801 Tungsten Road

Cleveland OH 44117-1117 USA

tel ++ 1 (216) 692 3990
fax ++ 1 (216) 692 0031
email
web address www.csm-moly.com

CSM is primarily an integrated source of molybdenum metal products. In addition, CSM offers extrusion services for a large variety of metals including metal matrix composites and can meet almost any specifications. (Contact Mr. Joe Hurt ++1 (517) 279 9511)

Update March 2003: Company name and web site have changed. New company name is H.C. Starck. Relevant link is automatic from web address above; information on extrusions, including metal matrix composite extrusions, can be found at
<http://www.hcstarck.com/index.php?page_id=194>

Cycotech Pty Ltd
40 Renowden Street
Cheltenham, Victoria 3192 Australia

tel ++ 61 3 9515 3678
fax ++ 61 3 9515 3740
email Cycotech@pro-race.com
web address www.Cyco-Systems.com

Cyco developed Ultalite, Aluminium reinforced with fly ash. Fly ash, a waste product from power stations, consists of ceramic spheres of about 10 to 50 micrometers diameter. Cyco uses white fly ash that consists primarily of Silica, Alumina and Titanium Dioxide. The material is low cost and is otherwise dumped to land fill. The density of the spheres is similar to that of Aluminium but they are also available in hollow form at about one quarter the density of Aluminium and are suitable for some applications.

Ultalite has been developed by Cyco as a joint program with the Australian Federal Government. The fly ash is mixed into the aluminium while it is heated in a custom designed mixing apparatus which ensures entrainment and wetting.

Ultalite was developed specifically for automobile brake components, for which a mass market exists. Ultalite benefits from its low cost, high modulus, light weight and abrasion resistance.
(information updated in Feb 2003)

Cymat Corporation

6320-2 Danville Road

Mississauga, Ontario L5T 2L7 Canada

tel ++ 1 (905) 696 9900
fax ++ 1 (905) 696 9300
email info@cymat.com
web address www.cymat.com

Cymat manufactures Stabilized Aluminum Foam (SAF), a technology that it licences from Alcan for manufacture and marketing worldwide. In addition, it has acquired the MMC foam production capacity of Norsk Hydro.

The structure of particulate-reinforced metals has distinct advantages for the production of metallic foams. The ceramic particles prevent the collapse of cell walls within the foam, thus significantly increasing the stability of closed-cell metal foams. Presently, Cymat markets foams made from silicon carbide and alumina particle-reinforced aluminium alloys, typically with reinforcement concentrations of 10 to 30 vol. %. This foam is manufactured by a continuous casting technique that was originally developed by Alcan; a similar method was developed by Norsk Hydro. The composite charge (usually a Duralcan derivative from Alcan) is first melted with conventional foundry equipment, then transferred to a tundish where gas is introduced to the melt producing the foam. The resulting liquid metal foam is carried away by conveyor belt; an upper belt is employed to create a flat face on the top surface of the foam.

Some of the benefits of Cymat's SAF include high specific strength, high specific energy absorption, good thermal and acoustic insulation, and lower cost than honeycomb. Various applications that benefit from this combination of properties are being developed: crash boxes, rail reinforcement, bumper systems, pillar reinforcement and seat structures for energy management and structural reinforcement. In addition to transportation-based applications, SAF potentially could be used for fire protection, blast protection for petrochemical plants and oil rigs and military armour.

Norsk Hydro (0240 Oslo Norway, www.hydro.com) owns part of Cymat.

DaimlerChrysler AG

Stuttgart 70546 Germany

tel ++ 49 (711) 170
fax ++ 49 (711) 179 4022
email
web address www.daimlerchrysler.com

DaimlerChrysler, like most major automobile manufacturers, has been engaged in development efforts that incorporate MMC and related technologies. Disc brakes have perhaps received the widest attention and these have been demonstrated on the rear axle of the Chrysler Prowler sports car. Advanced Metallic Composites produces these discs for Chrysler from Duralcan material (from Alcan).

In addition, Mercedes Benz (a DaimlerChrysler company), uses the Silitec technology, licenced from Osprey Metals, for cylinder liners. Although, Silitec is a hypereutectic aluminium-silicon alloy and not a MMC, it is related in that the final microstructure consists of primary silicon in an aluminium matrix.

DaimlerChrysler Corporation
Auburn Hills MI 48326
USA
tel ++1 (248) 576 5741
fax ++1 (248) 576 4742
www.chrysler.com

Dana Corporation

World Center of Driveshaft Technology
6201 Trust Drive
Holland OH 43528 USA

tel ++ 1 (419) 886 2600
fax ++ 1 (419) 535 4643
email
web address www.dana.com

Dana Corporation is a major supplier of components and systems for the automotive industry, with sales of over 10 billion USD in 2001. Spicer Met Com™ tubing (MMC tubing) is one of a range of materials developed by Dana's Automotive Systems Group to meet a growing range of driveline needs. This composite tubing is particularly well suited for high-performance driveshafts.

Delphi Automotive Systems
5725 Delphi Drive

Troy MI 48098-2815 USA

tel ++ 1 (248) 813 2000
fax ++ 1 (248) 813 2670
email
web address www.delphiauto.com

Delphi is a world leader in automotive systems technology. General Motors chose Delphi's Galileo™ braking system for the EV-1 (electric vehicle) with rear brake drums cast in particle-reinforced aluminium. The Galileo™ brake-by-wire system is a closed-loop, electrically activated system integrated to provide the braking functions.

dmc2 Electronic Components, now Ixion Thermal Materials

511 Manufacturers Rd.Chattanooga, TN 37405, USA
Sales: 1300 Marrows Rd.
Newark DE 19711 USA

tel ++ 1 (302) 456-6300
fax ++ 1 (302) 456-6333
email sales@ixiontm.com
web address http://www.ixiontm.com/

dmc2 Electronic Components (DEC) owns the intellectual property of the former Lanxide Electronic Components, Inc (LEC), including technology licences and manufacturing equipment. DEC is owned by dmc2 AG of the DHB Capital Group. DEC manufactures thermal management and structural products from the Al-SiC family of materials using the PRIMEX™ and PRIMEX Cast™ technologies developed by the former Lanxide Corporation. The PRIMEX™ pressureless infiltration technique produces components with between 50 and 70 vol. % SiC . The PRIMEX Cast™ process is used for components that require less reinforcement, 20 to 40 vol. % SiC. Thermal management solutions provided by DEC fall into two categories. Low to moderate thermal dissipation needs, typically achieved by direct-contact cooling, are called PRIMECOOL™, while the requirements of high heat removal are achieved with PRIMEFLO™ components that incorporate active cooling media. DEC provides extensive design guidance for machining, surface treatments, and metallization of its products. PRIMECOOL™ products have been used extensively in advanced engineering systems, particularly for power conversion baseplates in IGBTs (Insulated Gate Bipolar Transistor); some examples include GM's EV-1 (electric vehicle), Toyota's hybrid Prius, the Shinkansen 700 railway in Japan. Military applications of PRIMECOOL™ and PRIMEFLO™ components are numerous and these components reportedly can be found on the Eurofighter Typhoon, EA-6B Prowler, F-18 Hornet and particularly on the US Air Force's F-22 Raptor in the heads-up-display panel, fly-by-wire controls and the power generation unit. DEC's MMC components can also be found in space: PRIMECOOL™ printed wiring board cores were selected by Motorola for the Iridium satellite system, and both the Mars Pathfinder and the Cassini spacecraft have PRIMECOOL™ components aboard.

Update March 2003: activities of dmc2 have been taken over in 2002 by Ixion Thermal Materials; the address above is updated.

dmc2 Electronic Components, Corporation (DEC)
material (company designation): MCX-724™

Material Al / SiC / 55p

matrix unspecified grade Al
reinforcement SiC *volume percent* 55
morphology ☒ p ☐ sf ☐ cf ☐ mf ☐ w *size (μm)*

Processing
primary Infiltration
specifics PRIMEX™ pressureless infiltration

secondary ☐ Casting ☐ Forging ☐ Surface Treatments
☐ Extrusion ☐ Machining ☐ Thermal Treatments

specifics

joining ☐ Brazing ☐ Mechanical ☐ Welding ☐ Other
recyclable ☐ Yes ☐ No

Properties are quoted from manufacturer as general indicators only.

Properties *isotropic* ☒ Yes ☐ No *tensile*
density (g cm^{-3}) 2.94 *yield strength (MPa)*
CTE (x10^{6} K^{-1}) 7.2 (218-400K) *ultimate strength (MPa)*
thermal cond. (Wm^{-1}K^{-1}) 165 *modulus (GPa)* 235
electrical cond. (%IACS) *elongation (%)*

Property Advantages ☒ Density ☐ Strength ☐ Wear
☐ Fatigue ☒ Thermal Expansion ☐ Weldability
☒ Stiffness ☒ Thermal Conductivity ☐ Other

Applications cold plates, chip carriers, package bases

Products

Sources Lanxide Electronic Components (1999).

dmc2 Electronic Components, Corporation (DEC)
material (company designation): MCX-686™

Material Al / SiC / p

matrix unspecified grade Al

reinforcement SiC *volume percent*

morphology ☒ p ☐ sf ☐ cf ☐ mf ☐ w *size (μm)*

Processing

primary Infiltration

specifics PRIMEX™ pressureless infiltration

secondary ☐ Casting ☐ Forging ☐ Surface Treatments ☐ Extrusion ☐ Machining ☐ Thermal Treatments

specifics

joining ☐ Brazing ☐ Mechanical ☐ Welding ☐ Other

recyclable ☐ Yes ☐ No

Properties are quoted from manufacturer as general indicators only.

Properties *isotropic* ☐ Yes ☐ No *tensile*

density ($g\ cm^{-3}$) 3.00 *yield strength (MPa)*

CTE ($x10^{6}\ K^{-1}$) 6.8 (218-400K) *ultimate strength (MPa)*

thermal cond. ($Wm^{-1}K^{-1}$) 175 *modulus (GPa)* 235

electrical cond. (%IACS) *elongation (%)*

Property Advantages ☒ Density ☐ Strength ☐ Wear ☐ Fatigue ☒ Thermal Expansion ☐ Weldability ☒ Stiffness ☒ Thermal Conductivity ☐ Other

Applications microprocessor lids and heat spreaders, chip carriers

Products

Sources Lanxide Electronic Components (1999).

dmc2 Electronic Components, Corporation (DEC)
material (company designation): MCX-693™

Material Al / SiC / 70p

matrix unspecified grade Al
reinforcement SiC *volume percent* 70
morphology ☒ p ☐ sf ☐ cf ☐ mf ☐ w *size (μm)*

Processing

primary Infiltration
specifics PRIMEX™ pressureless infiltration

secondary ☐ Casting ☐ Forging ☐ Surface Treatments ☐ Extrusion ☐ Machining ☐ Thermal Treatments

specifics

joining ☐ Brazing ☐ Mechanical ☐ Welding ☐ Other
recyclable ☐ Yes ☐ No

Properties are quoted from manufacturer as general indicators only.

Properties *isotropic* ☐ Yes ☐ No *tensile*

density ($g\ cm^{-3}$) 2.98 *yield strength (MPa)*
CTE ($x10^6\ K^{-1}$) 6.4 (218-400K) *ultimate strength (MPa)*
thermal cond. ($Wm^{-1}K^{-1}$) 180 *modulus (GPa)* 225
electrical cond. (%IACS) *elongation (%)*

Property Advantages ☒ Density ☐ Strength ☐ Wear ☐ Fatigue ☒ Thermal Expansion ☐ Weldability ☒ Stiffness ☒ Thermal Conductivity ☐ Other

Applications printed wiring board cores, cold plates, chip carriers

Products

Sources Lanxide Electronic Components (1999).

dmc2 Electronic Components, Corporation (DEC)
material (company designation): MCX-696™

Material Al / SiC / p

matrix unspecified grade Al
reinforcement SiC *volume percent*
morphology ☒ p ☐ sf ☐ cf ☐ mf ☐ w *size (μm)*

Processing

primary Infiltration
specifics PRIMEX™ pressureless infiltration

secondary ☐ Casting ☐ Forging ☐ Surface Treatments ☐ Extrusion ☐ Machining ☐ Thermal Treatments

specifics

joining ☐ Brazing ☐ Mechanical ☐ Welding ☐ Other
recyclable ☐ Yes ☐ No

Properties are quoted from manufacturer as general indicators only.

Properties *isotropic* ☐ Yes ☐ No *tensile*

density (g cm^{-3}*)* 3.01 *yield strength (MPa)*
CTE (x10^{6} *K*$^{-1}$*)* 6.9 (218-400K) *ultimate strength (MPa)*
thermal cond. (Wm^{-1}*K*$^{-1}$*)* 192 *modulus (GPa)* 262
electrical cond. (%IACS) *elongation (%)*

Property Advantages ☒ Density ☐ Strength ☐ Wear ☐ Fatigue ☒ Thermal Expansion ☐ Weldability ☒ Stiffness ☒ Thermal Conductivity ☐ Other

Applications power module base plates, power amplifier heatsinks, microprocessor lids and heat spreaders, printed wiring board cores, cold plates, chip carriers

Products

Sources Lanxide Electronic Components (1999).

dmc2 Electronic Components, Corporation (DEC)
material (company designation): MCX-947™

Material Al / SiC / p

matrix unspecified grade Al
reinforcement SiC *volume percent*
morphology ☒ p ☐ sf ☐ cf ☐ mf ☐ w *size (μm)*

Processing
primary Infiltration
specifics PRIMEX™ pressureless infiltration

secondary ☐ Casting ☐ Forging ☐ Surface Treatments ☐ Extrusion ☐ Machining ☐ Thermal Treatments

specifics

joining ☐ Brazing ☐ Mechanical ☐ Welding ☐ Other
recyclable ☐ Yes ☐ No

Properties are quoted from manufacturer as general indicators only.

Properties *isotropic* ☐ Yes ☐ No *tensile*
density (g cm^{-3}) 2.95 *yield strength (MPa)*
CTE (x10^{6} K^{-1}) 9.4 (218-398K) *ultimate strength (MPa)*
thermal cond. (Wm^{-1}K^{-1}) 179 *modulus (GPa)* 196
electrical cond. (%IACS) *elongation (%)*

Property Advantages ☒ Density ☐ Strength ☐ Wear ☐ Fatigue ☒ Thermal Expansion ☐ Weldability ☒ Stiffness ☒ Thermal Conductivity ☐ Other

Applications thermal management in electronics

Products

Sources

dmc2 Electronic Components, Corporation (DEC)
material (company designation): MCX-1195™

Material Al / SiC / p

matrix unspecified grade Al
reinforcement SiC *volume percent*
morphology ☒ p ☐ sf ☐ cf ☐ mf ☐ w *size (μm)*

Processing

primary Infiltration
specifics PRIMEX™ pressureless infiltration

secondary ☒ Casting ☐ Forging ☐ Surface Treatments ☐ Extrusion ☐ Machining ☐ Thermal Treatments

specifics PRIMEX Cast™ process

joining ☐ Brazing ☐ Mechanical ☐ Welding ☐ Other
recyclable ☐ Yes ☐ No

Properties are quoted from manufacturer as general indicators only.

Properties *isotropic* ☐ Yes ☐ No *tensile*

density (g cm⁻³) 2.87 *yield strength (MPa)*
CTE (x10⁶ K⁻¹) 11.9 (218-398K) *ultimate strength (MPa)*
thermal cond. (Wm⁻¹K⁻¹) 183 *modulus (GPa)* 151
electrical cond. (%IACS) *elongation (%)*

Property Advantages ☒ Density ☐ Strength ☐ Wear ☐ Fatigue ☒ Thermal Expansion ☐ Weldability ☒ Stiffness ☒ Thermal Conductivity ☐ Other

Applications thermal management in electronics

Products

Sources

dmc2 Electronic Components, Corporation (DEC)
material (company designation): MCX-1405™

Material Al / SiC / p

matrix unspecified grade Al

reinforcement SiC *volume percent*

morphology ☒ p ☐ sf ☐ cf ☐ mf ☐ w *size (μm)*

Processing

primary Infiltration

specifics PRIMEX™ pressureless infiltration

secondary ☒ Casting ☐ Forging ☐ Surface Treatments ☐ Extrusion ☐ Machining ☐ Thermal Treatments

specifics PRIMEX Cast™

joining ☐ Brazing ☐ Mechanical ☐ Welding ☐ Other

recyclable ☐ Yes ☐ No

Properties are quoted from manufacturer as general indicators only.

Properties *isotropic* ☐ Yes ☐ No *tensile*

density (g cm⁻³) 2.80 *yield strength (MPa)*

CTE (x10⁶ K⁻¹) 14.0 (218-398K) *ultimate strength (MPa)*

thermal cond. (Wm⁻¹K⁻¹) 165 *modulus (GPa)* 125

electrical cond. (%IACS) *elongation (%)*

Property Advantages ☒ Density ☐ Strength ☐ Wear ☐ Fatigue ☒ Thermal Expansion ☐ Weldability ☒ Stiffness ☒ Thermal Conductivity ☐ Other

Applications thermal management in electronics

Products

Sources

dmc2 Electronic Components, Corporation (DEC)
material (company designation): MCX-1605™

Material Al / SiC / p

matrix unspecified grade Al
reinforcement SiC *volume percent*
morphology ☒ p ☐ sf ☐ cf ☐ mf ☐ w *size (μm)*

Processing
primary Infiltration
specifics PRIMEX™ pressureless infiltration

secondary ☒ Casting ☐ Forging ☐ Surface Treatments ☐ Extrusion ☐ Machining ☐ Thermal Treatments
specifics PRIMEX Cast™

joining ☐ Brazing ☐ Mechanical ☐ Welding ☐ Other
recyclable ☐ Yes ☐ No

Properties are quoted from manufacturer as general indicators only.

Properties *isotropic* ☐ Yes ☐ No *tensile*
density ($g\ cm^{-3}$) 2.77 *yield strength (MPa)*
CTE ($x10^6\ K^{-1}$) 16.2 (218-398K) *ultimate strength (MPa)*
thermal cond. ($Wm^{-1}K^{-1}$) 132 *modulus (GPa)* 106
electrical cond. (%IACS) *elongation (%)*

Property Advantages ☒ Density ☐ Fatigue ☒ Stiffness ☒ Strength ☒ Thermal Expansion ☒ Thermal Conductivity ☐ Wear ☐ Weldability ☐ Other

Applications thermal management in electronics

Products

Sources

DWA Aluminum Composites

21130 Superior Street

Chatsworth CA 91311 USA

tel ++ 1 (818) 998 1504
fax ++ 1 (818) 998 0343
email mvdbergh@dwa-dra.com
web address www.dwa-dra.com

DWA Aluminum Composites manufactures discontinuously reinforced aluminium (commonly referred to as DRA) by vacuum hot pressing powders into billets weighing up to 400 kg. These billets are then formed into a variety of products including extruded tube, rolled sheet and forgings.

DWA's standard structural products are based on 2009 and 6092 aluminium alloys reinforced with between 15 and 25 vol. % silicon carbide. Particulate concentrations between 5 and 55 vol. % are also produced by DWA, such as 6063/SiC/50p for thermal management in electronic packaging. Other composite combinations are available such as 7000 series aluminium matrix for high-strength applications, or boron carbide reinforcement for applications that require higher specific strength, and improved weldability. Due to high neutron shielding properties of boron, MMC with boron carbide are particularly attractive for containment of spent nuclear fuel.

DWA's discontinuously reinforced aluminium products meet the stringent requirements of the aerospace industry, and they have been used in a number of aerospace applications. Fan Exit Guide Vanes (FEGVs) on the Pratt & Whitney 4000 series aircraft engines for the Boeing 777, for example, are extruded 6092/SiC/17.5p MMC. Military applications include ventral fins and fuel access covers on the US Air Force F-16, and hydraulic components on the F-18E/F. Applications with high concentrations of reinforcement, 6063/SiC/50p, include electronic packaging chip carriers for Global Positioning System (GPS) satellites.

Titanium blade sleeves have been replaced by blade sleeves forged from 2009/SiC/15p in the Dauphin AS365 N4 and the EC 120 helicopters from Eurocopter. The MMC offers higher specific properties and thus lower component weight compared to titanium. Blade sleeves are an important first for DWA in particular and MMC in general, as this is the first discontinuously reinforced aluminium employed for Class 1 vital critical rotating parts. This component is forged by Forges de Bologne for Eurocopter.

DWA Aluminum Composites

material (company designation): AA2009 / SiC / 15.5p / extrusion /

Material 2009 / SiC / 15.5p / extrusion / T42

matrix Al alloy

reinforcement F1000 grit powder SiC *volume percent* 15.5

morphology ☒ p ☐ sf ☐ cf ☐ mf ☐ w *size (μm)* 5

Processing

primary Powder Metallurgy

specifics vacuum hot pressing

secondary ☐ Casting ☒ Forging ☐ Surface Treatments ☒ Extrusion ☐ Machining ☒ Thermal Treatments

specifics

joining ☐ Brazing ☐ Mechanical ☐ Welding ☐ Other

recyclable ☐ Yes ☐ No

Properties are quoted from manufacturer as general indicators only.

Properties *isotropic* ☐ Yes ☒ No

density (g cm^{-3})	2.82		*tensile*
		yield strength (MPa)	370
CTE (x10^6 K^{-1})		*ultimate strength (MPa)*	550
thermal cond. (Wm^{-1}K^{-1})		*modulus (GPa)*	96
electrical cond. (%IACS)		*elongation (%)*	7

Property Advantages ☒ Density ☒ Fatigue ☒ Stiffness ☐ Strength ☒ Thermal Expansion ☒ Thermal Conductivity ☒ Wear ☐ Weldability ☐ Other

Applications spacecraft struts and fittings, robotics, bicycle frames, hydraulic components and to replace titanium in rotating aerospace components

Products F/A-18E/F Super Hornet; hydraulic components; Dauphin and Colibri; blade sleeves

Sources DWA Aluminum Composites (2000).

DWA Aluminum Composites

material (company designation): AA6092 / SiC / 17.5p / extrusion /

Material 6092 / SiC / 17.5p / extrusion / T6

matrix Al alloy

reinforcement SiC *volume percent* 17.5

morphology ☒ p ☐ sf ☐ cf ☐ mf ☐ w *size (μm)*

Processing

primary Powder Metallurgy

specifics vacuum hot pressing

secondary ☐ Casting ☒ Forging ☐ Surface Treatments ☒ Extrusion ☐ Machining ☒ Thermal Treatments

specifics

joining ☐ Brazing ☐ Mechanical ☐ Welding ☐ Other

recyclable ☐ Yes ☐ No

Properties are quoted from manufacturer as general indicators only.

Properties *isotropic* ☐ Yes ☒ No

density (g cm^{-3})	2.79	*tensile yield strength (MPa)*	451
CTE (x10^{6} K^{-1})		*ultimate strength (MPa)*	514
thermal cond. (Wm^{-1}K^{-1})		*modulus (GPa)*	106
electrical cond. (%IACS)		*elongation (%)*	7

Property Advantages ☒ Density ☐ Strength ☒ Wear ☒ Fatigue ☒ Thermal Expansion ☐ Weldability ☐ Stiffness ☒ Thermal Conductivity ☐ Other

Applications specific stiffness in aerospace structures: guide vanes on gas turbine engines; braking applications on roller coasters; to be included in MIL-HDBK-5

Products PW4000 series aircraft Fan Exit Guide Vanes

Walt Disney World Big brake fins

Sources DWA Aluminum Composites (2000).

DWA Aluminum Composites

material (company designation): AA6092 / SiC / 25p / extrusion / T6

Material 6092 / SiC / 25p / extrusion / T6

matrix Al alloy

reinforcement SiC *volume percent* 25

morphology ☒ p ☐ sf ☐ cf ☐ mf ☐ w *size (μm)*

Processing

primary Powder Metallurgy

specifics vacuum hot pressing

secondary ☐ Casting ☒ Forging ☐ Surface Treatments ☒ Extrusion ☐ Machining ☒ Thermal Treatments

specifics

joining ☐ Brazing ☐ Mechanical ☐ Welding ☐ Other

recyclable ☐ Yes ☐ No

Properties are quoted from manufacturer as general indicators only.

Properties *isotropic* ☐ Yes ☒ No

			tensile
density (g cm^{-3})	2.82	*yield strength (MPa)*	420
CTE (x10^{6} K^{-1})	15.3	*ultimate strength (MPa)*	520
thermal cond. (Wm^{-1}K^{-1})		*modulus (GPa)*	115
electrical cond. (%IACS)		*elongation (%)*	5

Property Advantages ☒ Density ☐ Strength ☒ Wear ☒ Fatigue ☒ Thermal Expansion ☐ Weldability ☐ Stiffness ☒ Thermal Conductivity ☐ Other

Applications to be included in MIL-HDBK-5

Products

Sources DWA Aluminum Composites (2000).

DWA Aluminum Composites

material (company designation): AA6063 / SiC / 50p

Material 6063 / SiC / 50p

matrix Al alloy

reinforcement F1000 grit powder SiC *volume percent* 50

morphology ☒ p ☐ sf ☐ cf ☐ mf ☐ w *size (μm)* 5

Processing

primary Powder Metallurgy

specifics vacuum hot pressing

secondary ☐ Casting ☐ Forging ☐ Surface Treatments ☐ Extrusion ☐ Machining ☐ Thermal Treatments

specifics

joining ☐ Brazing ☐ Mechanical ☐ Welding ☐ Other

recyclable ☐ Yes ☐ No

Properties are quoted from manufacturer as general indicators only.

Properties *isotropic* ☐ Yes ☒ No *tensile*

density ($g\ cm^{-3}$) *yield strength (MPa)*

CTE ($x10^6\ K^{-1}$) *ultimate strength (MPa)*

thermal cond. ($Wm^{-1}K^{-1}$) *modulus (GPa)*

electrical cond. (%IACS) *elongation (%)*

Property Advantages ☒ Density ☐ Strength ☐ Wear ☐ Fatigue ☒ Thermal Expansion ☐ Weldability ☐ Stiffness ☒ Thermal Conductivity ☐ Other

Applications chip carriers in electronic packaging, particularly for satellites

Products Navstar GPS satellites electronic packaging

Sources DWA Aluminum Composites (2000).

DWA Aluminum Composites

material (company designation): AA6092 / SiC / 17.5p / sheet / T6

Material 6092 / SiC / 17.5p / sheet / T6

matrix Al alloy

reinforcement SiC *volume percent* 17.5

morphology ☒ p ☐ sf ☐ cf ☐ mf ☐ w *size (μm)*

Processing

primary Powder Metallurgy

specifics vacuum hot pressing

secondary ☐ Casting ☒ Forging ☐ Surface Treatments ☒ Extrusion ☐ Machining ☒ Thermal Treatments

specifics

joining ☐ Brazing ☐ Mechanical ☐ Welding ☐ Other

recyclable ☐ Yes ☐ No

Properties are quoted from manufacturer as general indicators only.

Properties *isotropic* ☐ Yes ☒ No

			tensile
density (g cm^{-3})	2.79	*yield strength (MPa)*	395
CTE (x10^{6} K^{-1})		*ultimate strength (MPa)*	460
thermal cond. (Wm^{-1}K^{-1})		*modulus (GPa)*	102
electrical cond. (%IACS)		*elongation (%)*	8

Property Advantages ☒ Density ☐ Strength ☒ Wear ☒ Fatigue ☒ Thermal Expansion ☐ Weldability ☐ Stiffness ☒ Thermal Conductivity ☐ Other

Applications high specific stiffness and strength in sheet form such as ventral fins for military aircraft; to be included in MIL-HDBK-5

Products F-16 Fighting Falcon ventral fins
F-16 Fighting Falcon fuel access covers

Sources DWA Aluminum Composites (2000).

Dynamet Technology, Inc
Eight A Street

Burlington MA 01803 USA

tel ++ 1 (781) 272 5967
fax ++ 1 (781) 229 4879
email info@dynamettechnology.com
web address www.dynamettechnology.com

Dynamet Technology manufactures two distinct types of MMC materials: titanium reinforced with TiC or TiB, and aluminium reinforced with B_4C. Both materials are manufactured by a powder metallurgical route called the CHIP process, Cold plus Hot Isostatic Pressing. In this process powders of the matrix alloy and the reinforcement are blended, then cold isostatic pressed and subsequently sintered. This sintered material is further consolidated by HIPing, extruding, forging or a combination thereof.

The titanium-based MMC are called CermeTi® and reinforced with TiC particulate, or "whisker-like" TiB particles formed by the addition of TiB_2 to the powder blend. These materials are tailored for specific applications with concentrations from 5 to 20 vol. %. These MMC exhibit enhanced wear resistance, hardness, stiffness and elevated temperature strength compared with unreinforced titanium alloys and the reinforcement does not compromise the intrinsic low density, corrosion resistance and nonmagnetic properties of the titanium alloy. Applications for these titanium composites are being developed and include shot sleeves for the aluminium die casting industry, automotive components such as connecting rods, pistons and valves, and a number of recreational products such as faces for golf club heads, blades for ice skates, and sporting knives.

The Al-B_4C composites produced by Dynamet Technology are used for nuclear shielding applications due to the high neutron attenuation of boron carbide.

Dynotech Engineering Services
1635 Northwood
Troy MI 48084-5302 USA

tel ++ 1 (248) 362 2777 // 1 (800) 633 5559
fax ++ 1 (248) 362 3886
email dynotech@dynotechengineering.com
web address www.dynotechengineering.com

Dynotech manufactures aftermarket driveshafts with emphasis on motorsport, aiming to improve delivery of power to the rear wheels. By virtue of their higher specific stiffness, MMC allows driveshafts to be operated at higher speeds before the rotation of the shaft becomes unstable or starts to wobble. The critical operation speed of an MMC driveshaft is 20% greater than a 6061-T6 aluminium driveshaft, with torque capability enhanced by 40%. Weight reductions of 60% are achieved as compared to steel driveshafts.

Edelstahl Witten Krefeld GmbH

Sonderwerkstoffe
Gladbacher Strasse 578
Krefeld 47805 Germany

tel ++ 49 (2151) 83 38 44
fax ++ 49 (2151) 83 38 77
email sales@ferro-titanit.com
web address www.ferro-titanit.com

Edelstahl Witten-Krefeld manufactures and markets a family of steel-matrix composites called Ferro-Titanit® for high wear applications. These materials contain as much as 45 vol. % titanium carbide embedded in a steel matrix, fabricated by a powder metallurgical route. Conventional machining techniques can be used in the soft annealed condition and Edelstahl Witten-Krefeld provides detailed machining recommendations. Thermal treatments yield a final hardness of up to 69 HRC with little distortion of the material due to the low thermal expansion and lack of phase change in the TiC phase. Additionally, Ferro-Titanit® can be "recycled" through annealing and re-machined into new parts.

Ferro-Titanit® is offered in seven grades to meet the requirements of various applications. The C-Spezial grade has a low-alloy steel matrix, thus it exhibits a low resistance to tempering and a decrease of hardness above 200°C; however, C-Spezial offers the best machining characteristics in the Ferro-Titanit® family. This material is best suited for general cold-work applications in cutting and forging. The WFN grade has comparably higher alloy content, providing tempering resistance up to 450°C in addition to high-temperature hardness and good corrosion resistance. WFN is generally recommended for cold cutting and forming, such as valve components, rollers for wire rod and bar steel rolling, and moulds for injection moulding of polymers. The S grade offers enhanced resistance to corrosion, as required in pump applications and thrust disks. The Nikro 128 and Nikro 143 grades have high nickel contents to produce a tough age-hardenable nickel martensite. Nikro 128 is suitable for processing of abrasive plastics and Nikro 143 is recommended for heavy wear and bending applications at elevated temperature. Cromoni is an austenitic grade for nonmagnetic applications requiring maximum wear and corrosion resistance. The U grade is an austenitic matrix that offers excellent corrosion resistance particularly in the presence of chlorine ions and is thus suited for the chemical industry.

Edelstahl Witten Krefeld GmbH

material (company designation): Ferro-Titanit™ C-Spezial

Material Fe / TiC / 43p

matrix Fe - 3Cr - 3Mo - 0.65C

reinforcement TiC *volume percent* 43

morphology ☒ p ☐ sf ☐ cf ☐ mf ☐ w *size (μm)*

Processing

primary Powder Metallurgy

specifics

secondary ☐ Casting ☐ Forging ☐ Surface Treatments ☐ Extrusion ☒ Machining ☒ Thermal Treatments

specifics

joining ☒ Brazing ☒ Mechanical ☐ Welding ☒ Other

recyclable ☒ Yes ☐ No

Properties are quoted from manufacturer as general indicators only.

Properties *isotropic* ☒ Yes ☐ No *tensile*

density ($g\ cm^{-3}$) 6.5 *yield strength (MPa)*

CTE ($x10^{6}\ K^{-1}$) 9.2 (20-100°C) *ultimate strength (MPa)*

thermal cond. ($Wm^{-1}K^{-1}$) 20.9 *modulus (GPa)* 292

electrical cond. (%IACS) 2.3 * *elongation (%)*

Property Advantages ☒ Density ☐ Strength ☒ Wear ☐ Fatigue ☐ Thermal Expansion ☐ Weldability ☐ Stiffness ☐ Thermal Conductivity ☐ Other

Applications for cold metalworking: cutting tools, deep-drawing dies, punches, bending jaws; also for heavy wear as in pulleys, rollers and guides
* 0.75 $\Omega.mm^2/m$

Products

Sources

Edelstahl Witten Krefeld GmbH

material (company designation): Ferro-Titanit™ Nikro 143

Material Fe / TiC / 41p

matrix Fe - 9Co - 15Ni - 6Mo

reinforcement TiC *volume percent* 41

morphology ☒ p ☐ sf ☐ cf ☐ mf ☐ w *size (μm)*

Processing

primary Powder Metallurgy

specifics

secondary ☐ Casting ☐ Forging ☐ Surface Treatments ☐ Extrusion ☒ Machining ☒ Thermal Treatments

specifics

joining ☒ Brazing ☒ Mechanical ☐ Welding ☒ Other

recyclable ☒ Yes ☐ No

Properties are quoted from manufacturer as general indicators only.

Properties *isotropic* ☐ Yes ☐ No *tensile*

density (g cm^{-3}) 6.7 *yield strength (MPa)*

CTE (x10^{6} K^{-1}) 8.0 (20-100°C) *ultimate strength (MPa)*

thermal cond. (Wm^{-1}K^{-1}) 18.5 *modulus (GPa)* 280

electrical cond. (%IACS) 2.13* *elongation (%)*

Property Advantages ☐ Density ☐ Strength ☒ Wear ☐ Fatigue ☐ Thermal Expansion ☐ Weldability ☐ Stiffness ☐ Thermal Conductivity ☐ Other

Applications general purpose for tooling exposed to heavy wear up to 500°C, particularly processing of plastics as pelletiser knives, extruder works
* 0.81 Ω.mm^2/m

Products

Sources

Edelstahl Witten Krefeld GmbH

material (company designation): Ferro-Titanit™ WFN

Material Fe / TiC / 43p

matrix Fe - 13.5Cr - 3Mo - 0.75C

reinforcement TiC *volume percent* 43

morphology ☒ p ☐ sf ☐ cf ☐ mf ☐ w *size (μm)*

Processing

primary Powder Metallurgy

specifics

secondary ☐ Casting ☐ Forging ☐ Surface Treatments
☐ Extrusion ☒ Machining ☒ Thermal Treatments

specifics

joining ☒ Brazing ☐ Mechanical ☐ Welding ☒ Other

recyclable ☒ Yes ☐ No

Properties are quoted from manufacturer as general indicators only.

Properties *isotropic* ☐ Yes ☐ No *tensile*

density (g cm^{-3}) 6.5

CTE (x10^{6} K^{-1}) 10.6 (20-100°C)

thermal cond. (Wm^{-1}K^{-1}) 18.2

electrical cond. (%IACS) 1.9*

yield strength (MPa)

ultimate strength (MPa)

modulus (GPa) 294

elongation (%)

Property Advantages ☒ Density ☐ Strength ☒ Wear
☐ Fatigue ☒ Thermal Expansion ☐ Weldability
☒ Stiffness ☐ Thermal Conductivity ☐ Other

Applications general purpose for cold working applications: guide rollers for wire and rod rolling, valve components, drawing and extrusion dies
* 0.91 Ω.mm^2/m

Products

Sources

Edelstahl Witten Krefeld GmbH

material (company designation): Ferro-Titanit™ Cromoni

Material Ni / TiC / 33p

matrix Ni - 20Cr - 15.5Mo

reinforcement TiC *volume percent* 33

morphology ☒ p ☐ sf ☐ cf ☐ mf ☐ w *size (μm)*

Processing

primary Powder Metallurgy

specifics

secondary ☐ Casting ☐ Forging ☐ Surface Treatments ☐ Extrusion ☒ Machining ☒ Thermal Treatments

specifics

joining ☒ Brazing ☒ Mechanical ☐ Welding ☒ Other

recyclable ☒ Yes ☐ No

Properties are quoted from manufacturer as general indicators only.

Properties *isotropic* ☐ Yes ☐ No *tensile*

density (g cm^{-3}) 7.4 *yield strength (MPa)*

CTE (x10^{6} K^{-1}) 9.0 (20-100°C) *ultimate strength (MPa)*

thermal cond. (Wm^{-1}K^{-1}) 12.4 *modulus (GPa)* 277

electrical cond. (%IACS) 1.13* *elongation (%)*

Property Advantages ☐ Density ☐ Strength ☒ Wear ☐ Fatigue ☐ Thermal Expansion ☐ Weldability ☐ Stiffness ☒ Thermal Conductivity ☒ Other

Applications for wear and corrosion resistance in non-magnetic applications

* 1.53 Ω.mm^2/m

Products

Sources

Edelstahl Witten Krefeld GmbH

material (company designation): Ferro-Titanit™ S

Material Fe / TiC / 42p

matrix Fe - 19.5Cr - 2Mo - 0.5C

reinforcement TiC *volume percent* 42

morphology ☒ p ☐ sf ☐ cf ☐ mf ☐ w *size (μm)*

Processing

primary Powder Metallurgy

specifics

secondary ☐ Casting ☐ Forging ☐ Surface Treatments ☐ Extrusion ☒ Machining ☒ Thermal Treatments

specifics

joining ☒ Brazing ☒ Mechanical ☐ Welding ☒ Other

recyclable ☒ Yes ☐ No

Properties are quoted from manufacturer as general indicators only.

Properties *isotropic* ☐ Yes ☐ No

density (g cm^{-3}) 6.5

CTE (x10^{6} K^{-1}) 7 (20-400°C)

thermal cond. (Wm^{-1}K^{-1}) 18.8

electrical cond. (%IACS) 2.24*

tensile

yield strength (MPa)

ultimate strength (MPa)

modulus (GPa) 290

elongation (%)

Property Advantages ☒ Density ☐ Strength ☒ Wear ☐ Fatigue ☐ Thermal Expansion ☐ Weldability ☒ Stiffness ☐ Thermal Conductivity ☒ Other

Applications for wear and corrosion resistance in pumps, bearings etc.
* 0.77 Ω.mm^2/m

Products

Sources

Edelstahl Witten Krefeld GmbH

material (company designation): Ferro-Titanit™ Nikro 128

Material Fe / TiC / 40p

matrix Fe - 13.5Cr - 9Co - 4Ni - 5Mo

reinforcement TiC *volume percent* 40

morphology ☒ p ☐ sf ☐ cf ☐ mf ☐ w *size (μm)*

Processing

primary Powder Metallurgy

specifics

secondary ☐ Casting ☐ Forging ☐ Surface Treatments ☐ Extrusion ☒ Machining ☒ Thermal Treatments

specifics

joining ☒ Brazing ☒ Mechanical ☐ Welding ☒ Other

recyclable ☒ Yes ☐ No

Properties are quoted from manufacturer as general indicators only.

Properties *isotropic* ☐ Yes ☐ No *tensile*

density (g cm^{-3}) 6.6

CTE (x10^{6} K^{-1}) 8.3 (20-100°C)

thermal cond. (Wm^{-1}K^{-1}) 21

electrical cond. (%IACS) 1.57*

yield strength (MPa)

ultimate strength (MPa)

modulus (GPa) 294

elongation (%)

Property Advantages ☐ Density ☐ Strength ☒ Wear ☐ Fatigue ☐ Thermal Expansion ☐ Weldability ☐ Stiffness ☐ Thermal Conductivity ☐ Other

Applications for processing of plastics: pelletiser knives, dies, worms, bushes; also for wear-resistant machine components
* 1.1 Ω.mm^2/m

Products

Sources

Edelstahl Witten Krefeld GmbH

material (company designation): Ferro-Titanit™ U

Material Fe / TiC / 45p

matrix Fe - 18Cr - 12Ni - 2Mo

reinforcement TiC *volume percent* 45

morphology ☒ p ☐ sf ☐ cf ☐ mf ☐ w *size (μm)*

Processing

primary Powder Metallurgy

specifics

secondary ☐ Casting ☐ Forging ☐ Surface Treatments ☐ Extrusion ☒ Machining ☒ Thermal Treatments

specifics

joining ☒ Brazing ☒ Mechanical ☐ Welding ☒ Other

recyclable ☒ Yes ☐ No

Properties are quoted from manufacturer as general indicators only.

Properties *isotropic* ☐ Yes ☐ No *tensile*

density (g cm^{-3}) 6.6 — *yield strength (MPa)*

CTE (x10^{6} K^{-1}) 12.5 (20-400°C) — *ultimate strength (MPa)*

thermal cond. (Wm^{-1}K^{-1}) 18.0 — *modulus (GPa)*

electrical cond. (%IACS) 1.8 — *elongation (%)*

Property Advantages ☐ Density ☐ Strength ☒ Wear ☐ Fatigue ☐ Thermal Expansion ☐ Weldability ☐ Stiffness ☐ Thermal Conductivity ☒ Other

Applications for wear and corrosion resistance in non-magnetic applications, especially in the chemical industry

* 0.96 Ω.mm^2/m

Products

Sources

Electrovac GmbH

Aufeldgasse 37-39

Klosterneuburg 3400 Austria

tel ++ 43 (2243) 450 0
fax ++ 43 (2243) 450 698
email verkauf@electrovac.com
web address www.electrovac.com

Electrovac has developed ELCOM ASC-500, a composite of aluminium reinforced with about 70 vol. % silicon carbide particles for electronic packaging. Silicon carbide preforms are infiltrated with molten aluminium to produce net-shape components with an envelope of unreinforced material on all surfaces of packaging materials to facilitate plating.

Electrovac's MMC provides packaging solutions for power and high frequency electronic components where thermal management is a prime concern. Packaging solutions that benefit from the properties of MMC include FET (Field Effect Transistors), IGBT (Insulated Gate Bipolar Transistors) and diodes for power electronics in automobiles, as well as radar systems, telecommunications and other high-frequency electronics. In addition, MMC can be employed in integrated circuit technology as lids and covers, and as frames and housings.

(updated Feb. 2003)

Electrovac GmbH

material (company designation): ELCOM ASC-500

Material Al / SiC / 70p

matrix Al - 7Si - 0.3Mg

reinforcement SiC *volume percent* 70

morphology ☒ p ☐ sf ☐ cf ☐ mf ☐ w *size (μm)*

Processing

primary Infiltration

specifics net-shape infiltration

secondary ☐ Casting ☐ Forging ☒ Surface Treatments ☐ Extrusion ☐ Machining ☐ Thermal Treatments

specifics

joining ☒ Brazing ☐ Mechanical ☐ Welding ☐ Other

recyclable ☐ Yes ☐ No

Properties are quoted from manufacturer as general indicators only.

Properties *isotropic* ☐ Yes ☐ No *tensile*

density ($g\ cm^{-3}$) 3.0 *yield strength (MPa)*

CTE ($x10^6\ K^{-1}$) 6-8 (20-300°C) *ultimate strength (MPa)*

thermal cond. ($Wm^{-1}K^{-1}$) 180-200 *modulus (GPa)* 200

electrical cond. (%IACS) *elongation (%)*

Property Advantages ☒ Density ☐ Strength ☐ Wear ☐ Fatigue ☒ Thermal Expansion ☐ Weldability ☐ Stiffness ☒ Thermal Conductivity ☒ Other

Applications power electronics (IGBT module packages, power diodes); hight frequency electronics (radar systems, telecommunications); IC components (lids and covers); rectifier packaging

Products

Sources

Enertron

Suite no. 5
100 W. Hoover
Mesa AZ 85210 USA

tel ++ 1 (480) 649 5400 // 1 (877) 363 7876
fax ++ 1 (480) 649 5434
email info@enertron-inc.com
web address www.enertron-inc.com

Enertron specialises in thermal management solutions for industries ranging from microelectronics to aerospace and defence. MMC materials are particularly recommended for packaging with high-power density. Enertron offers solutions from both aluminium- and copper-based MMC that provide good thermal conductivity with the ability to modify the concentration of reinforcement so as to match the thermal expansion of the composite to silicon, alumina or GaAs.

Eurocopter

European Aerospace Defence and Space Company (EADS)
Aéroport International Marseille Provence
Marignane 13725 France

tel ++ 33 (442) 85 95 55
fax ++ 33 (442) 85 95 64
email
web address www.eurocopter.com (www.eads.net)

Eurocopter has adopted particle-reinforced aluminium to replace titanium for blade sleeves in two models of helicopter, the Dauphin AS365 N4 and Colibri EC 120. The blade sleeves connect the rotors to the rotor hub and are produced by Forges de Bologne from 2009/SiC/15p supplied by DWA Aluminum Composites. The MMC component is less expensive than the titanium solution and this application represents the first time that discontinuously reinforced aluminium has been used for a safety-critical, rotating aerospace component. The swash plate assembly in the EC 135 helicopter is also partially MMC, produced from 2009/SiC/15p supplied by DWA; the swash plate assembly controls the lift and orientation of the helicopter. These are civilian helicopters.

Eurocopter is a wholly owned subsidiary of the European Aerospace Defence and Space Company (EADS). EADS is the product of the multinational merger of DaimlerChrysler Aerospace, Aerospatiale Matra and CASA (formerly the Spanish aerospace company).

Feinguss-Blank

Industriestrasse 18

Riedlingen 88499 Germany

tel ++ 49 (7371) 182 0
fax ++ 49 (7371) 13011
email fgblank@feinguss-blank.de
web address www.feinguss-blank.de

Feinguss-Blank is an investment casting company with expertise in producing continuous fibre reinforced aluminium castings. This technology is a spin-off of a multiyear research effort at RWTH in Aachen; thus, it can be inferred that a low-pressure gas-driven molten-aluminium infiltration process is used. The matrix alloy is specially designed and based on 7000 series aluminium alloys, aiming to combine attractive properties in both the unreinforced and reinforced portions of selectively reinforced castings. The reinforcement is typically continuous alumina fibres and these can be strategically positioned within components of complex shapes. The principal aim of selective reinforcement in investment castings is to increase the strength and stiffness at the most stressed areas of the component; continuous alumina fibres increase the strength of the material parallel to the fibre axis by a factor of two.

It remains unclear if Feinguss-Blank has realised commercial production of its selectively reinforced investment castings, such as bicycle cranks, or whether the technology awaits exploitation.

Note added in Feb. 2003: since this database was produced, Feinguss-blank has stopped all activity in this area and is no longer active in fibre reinforced metals.

Feinguss-Blank

material (company designation):

Material Al / Al2O3 / cf

matrix Al-Zn-Mg

reinforcement Altex fibres Al2O3 *volume percent*

morphology ☐ p ☐ sf ☒ cf ☐ mf ☐ w *size (μm)*

Processing

primary Infiltration

specifics investment casting

secondary ☐ Casting ☐ Forging ☐ Surface Treatments ☐ Extrusion ☐ Machining ☐ Thermal Treatments

specifics

joining ☐ Brazing ☐ Mechanical ☐ Welding ☐ Other

recyclable ☐ Yes ☐ No

Properties are quoted from manufacturer as general indicators only.

Properties *isotropic* ☐ Yes ☒ No

density $(g\ cm^{-3})$		*tensile* *yield strength (MPa)*	
CTE $(x10^{6}\ K^{-1})$	5.6	*ultimate strength (MPa)*	900
thermal cond. $(Wm^{-1}K^{-1})$		*modulus (GPa)*	135
electrical cond. (%IACS)		*elongation (%)*	

Property Advantages ☐ Density ☒ Strength ☐ Wear ☐ Fatigue ☐ Thermal Expansion ☐ Weldability ☒ Stiffness ☐ Thermal Conductivity ☐ Other

Applications

Products

Sources

FiberNide Ltd

3217 Shoreline Drive

Oakville, Ontario L6L5Y9 Canada

tel ++ 1 (905) 330 1239
fax ++
email info@fibernide.com
web address www.fibernide.com

FiberNide was established to commercialise technology advances in the production of nickel aluminide materials, particularly nickel aluminide-based composites and foams. The FiberNide™ material is a nickel aluminide reinforced with continuous carbon fibres. Carbon fibre tows (containing thousands of individual fibres) are coated with a layer of nickel and subsequently with a layer of aluminium: both layers are produced by Chemical Vapour Deposition (CVD) using organo-metallic precursors. For consolidation, the bundles of coated fibres are compressed in a graphite die under vacuum at 1200°C. The nickel aluminide is formed by exothermic reaction between the nickel and aluminium during processing.

FiberNide™ is typically reinforced with 50 vol. % carbon fibres and the matrix phase can be varied to have compositions from NiAl to Ni_3Al. At temperatures between 400-700°C FiberNide™ reportedly exhibits the highest specific strength of any aerospace material, hence the interest in this material for aerospace structures and turbine applications. FiberNide supplies both consolidated MMC and loose "coated-but-unreacted" fibre bundles. Future developments may include a silicon carbide reinforced nickel aluminide.

FiberNide Ltd

material (company designation):

Material NiAl-Ni3Al / carbon / 50cf

matrix Ni-Al

reinforcement carbon fibre *volume percent* 50

morphology ☐ p ☐ sf ☒ cf ☐ mf ☐ w *size (μm)* 7

Processing

primary Deposition

specifics Chemical Vapour Deposition (CVD) using

secondary ☐ Casting ☐ Forging ☐ Surface Treatments ☐ Extrusion ☐ Machining ☐ Thermal Treatments

specifics diffusion bonding

joining ☐ Brazing ☐ Mechanical ☐ Welding ☐ Other

recyclable ☐ Yes ☐ No

Properties are quoted from manufacturer as general indicators only.

Properties *isotropic* ☐ Yes ☐ No *tensile*

density (g cm^{-3}) 3.9 *yield strength (MPa)* 800

CTE (x10^{6} K^{-1}) *ultimate strength (MPa)*

thermal cond. (Wm^{-1}K^{-1}) *modulus (GPa)* 115

electrical cond. (%IACS) *elongation (%)*

Property Advantages ☐ Density ☒ Strength ☐ Wear ☐ Fatigue ☐ Thermal Expansion ☐ Weldability ☐ Stiffness ☐ Thermal Conductivity ☐ Other

Applications specific strength at temperatures between 400-700°C

Products

Sources

Ford Motor Company

Dearborn MI 48126 USA

tel ++
fax ++
email
web address www.ford.com

It was reported that Ford has been actively engaged in the development of MMC brake disks (silicon carbide particle reinforced aluminium), but this application has not (yet?) been implemented. Aluminium-matrix composite driveshafts, however, have been implemented by Ford in the Police Interceptor model of the Crown Victoria; this application represents 65,000 units annually. This lightweight driveshaft is manufactured by Ford's subsidiary Visteon from seamless extruded and drawn MMC tube purchased from Alcoa. The alumina-particle reinforced aluminium originates from Alcan (Duralcan material). Due to its high specific stiffness, MMC driveshafts can be run at higher speeds than steel or aluminium driveshafts.

In addition to this application, Ford will reportedly employ MMC brake discs on their forthcoming hybrid vehicle, the Ford Prodigy, a hybrid electric family car prototype that can achieve a fuel economy of 70 mpg (miles per gallon). The discs are to be manufactured from the Duralcan material A359/SiC/20p (from Alcan).

(this information is from a news site on the web; it was not confirmed by direct contact with Ford researchers in Feb. 2003)

Forges de Bologne

39, rue des Forges

Bologne 52310 France

tel ++ 33 (325) 32 41 80
fax ++ 33 (325) 32 39 21 / 80 07
email
web address

Forges de Bologne manufactures forged components that are used in a wide variety of aeronautical applications, including MMC helicopter blade sleeves for the Eurocopter Dauphin AS365 N4 and the Colibri EC 120. The blade sleeves are forged from 2009/SiC/15p supplied by DWA Aluminum Composites. This material offers significant improvements in fatigue resistance compared to conventional heat-treatable aluminium alloys: 250 MPa compared to 180 MPa at 10^7 cycles of rotating bending fatigue. This MMC replaced titanium since it displays higher specific properties and thus lower weight for the blade sleeve. In addition to blade sleeves, Forges de Bologne forges part of the swash plate assembly from the same MMC for Eurocopter's EC 135 helicopter.

Forges de Bologne
material (company designation):

Material 2080 / SiC / 15p / T4

matrix Al alloy

reinforcement SiC *volume percent* 15

morphology ☒ p ☐ sf ☐ cf ☐ mf ☐ w *size (μm)* 8-10

Processing

primary

specifics

secondary ☐ Casting ☒ Forging ☐ Surface Treatments ☐ Extrusion ☒ Machining ☐ Thermal Treatments

specifics

joining ☐ Brazing ☐ Mechanical ☐ Welding ☐ Other

recyclable ☐ Yes ☐ No

Properties are quoted from manufacturer as general indicators only.

Properties *isotropic* ☐ Yes ☐ No

			tensile
density ($g\ cm^{-3}$)	2.83	*yield strength (MPa)*	370
CTE ($x10^6\ K^{-1}$)		*ultimate strength (MPa)*	480
thermal cond. ($Wm^{-1}K^{-1}$)		*modulus (GPa)*	100
electrical cond. (%IACS)		*elongation (%)*	7.5

Property Advantages ☐ Density ☐ Strength ☐ Wear ☒ Fatigue ☒ Thermal Expansion ☐ Weldability ☒ Stiffness ☐ Thermal Conductivity ☐ Other

Applications for replacement of titanium structural components: helicopter rotor control spiders and swashplates, stringers, missile components, connecting rods, pistons

Products

Sources

Forges de Bologne

material (company designation):

Material 2080 / SiC / 25p / T4

matrix Al alloy

reinforcement SiC *volume percent* 25

morphology ☒ p ☐ sf ☐ cf ☐ mf ☐ w *size (μm)* 8-10

Processing

primary

specifics

secondary ☐ Casting ☒ Forging ☐ Surface Treatments ☐ Extrusion ☒ Machining ☐ Thermal Treatments

specifics

joining ☐ Brazing ☐ Mechanical ☐ Welding ☐ Other

recyclable ☐ Yes ☐ No

Properties are quoted from manufacturer as general indicators only.

Properties *isotropic* ☐ Yes ☐ No

density (g cm^{-3})	2.83		*tensile*
CTE (x10^{6} K^{-1})		*yield strength (MPa)*	370
thermal cond. (Wm^{-1}K^{-1})		*ultimate strength (MPa)*	480
electrical cond. (%IACS)		*modulus (GPa)*	100
		elongation (%)	7.5

Property Advantages ☐ Density ☐ Strength ☐ Wear ☒ Fatigue ☒ Thermal Expansion ☐ Weldability ☒ Stiffness ☐ Thermal Conductivity ☐ Other

Applications for replacement of titanium structural components: helicopter rotor control spiders and swashplates, stringers, missile components, connecting rods, pistons

Products

Sources

Foster-Miller Inc
303 Bear Hill Road,

Waltham MA 02451 USA

tel ++ 1 (781) 684-4125
fax ++ 1 (781) 890-0488
email ukashalikar@foster-miller.com
web address www.foster-miller.com

Foster-Miller is a diverse engineering company that serves a number of industrial sectors; MMC activities at Foster-Miller represent a small fraction of its development and manufacturing ability. Under the NASA SBIR/STTR programs, Foster-Miller developed and manufactured a lightweight radiator for applications in space using a composite of aluminium and graphite (Al-Gr). This structural radiator removes heat from electronics and has generated over 40,000 USD in sales. Foster-Miller has also used this composite material for other components including dimensionally stable structures for space, optical platforms and for thermal management in gas turbines and rocket engines. This material is also suitable as a substrate for thermal management of dense electronic packages as in radar and satellites among others.

The Small Business Innovation Research (SBIR) and the Small Business Technology Transfer (STTR) programmes are funded by the US government to assist small businesses in expanding research and development.

Foster-Miller is also developing low cost particulate reinforced aluminum MMCs with high strength and stiffness, and property retention at temperatures up to 500 F. This material is suitable for applications in diesel engines and tribological components.

Separately, Foster-Miller is also developing attachment techniques for MMCs, including selective reinforcement of aluminum gravity castings.

Since 2001, Foster-Miller is producing an isotropic particulate reinforced alumina/aluminum MMC material with a tensile strength of 700 MPa (100 Ksi) and modulus of 210 GPa (30 Msi). At 260 C (500 F), the MMC retains 85 percent of room temperature strength and 95 percent of room temperature stiffness. This MMC material has also been used for selective reinforcement of aluminum gravity castings. *(information updated in Feb 2003)*

Fox Golf

45 E. Herbert Avenue

Salt Lake City UT 84111 USA

tel ++	1 (801) 364 1332 // 1 (800) 752 9369
fax ++	1 (801) 364 1363
email	sales@playfox.com
web address	www.playfox.com

Fox Golf manufactures the Uno® 2K drivers and woods with MMC heads. This head is larger, lighter and stronger than conventional club heads. The specifications of the MMC that is employed was not found, although from the benefits cited by Fox Golf it is suspected that the MMC is a light alloy reinforced with ceramic particulate.

General Motors (see web site)

USA

tel ++
fax ++
email
web address www.gm.com

In 1997 General Motors (GM) began commercially exploiting MMC materials by introducing a MMC engine cradle and an extruded particle-reinforced aluminium propshaft on the Corvette automobile. The propshaft was manufactured by Unidrive, an Australian supplier to the automotive industry that obtained extruded Duralcan material (from Alcan) from Capral, an Australian extrusion house. The high specific stiffness of the MMC propshaft allows the engine speed to be increased to 6200 rpm. Unfortunately, the increased cost of the composite driveline system was not sustainable and Unidrive now supplies an all-aluminium (6061) propshaft for the Corvette.

Also in 1997, GM introduced a MMC driveshaft for the extended S/T truck as a means of reducing weight and increasing performance. Due to the extended wheelbase, the driveshaft for this S/T is longer. The performance of conventional steel driveshafts is limited by the specific stiffness of steel, thus a simple extension of the driveshaft would result in rotational instabilities. Aluminium-matrix composites, however, have a higher specific stiffness than steel, allowing one-piece driveshafts that do not display instabilities in service. The aluminium composite driveshaft results in a 50% reduction in weight compared to a two-piece steel shaft connected by a universal joint, the previous engineering solution. In addition, the extrusion of the composite (Al/Al2O3/20p) is 20% less expensive. This application is believed to be the highest volume application of MMC to date with approximately 150,000 units produced annually.

These innovations are complemented by additional applications of MMC materials in the revolutionary EV-1 (electric vehicle), reportedly containing as many as 12. These include the well-publicised MMC rear brake drums in the Galileo™ braking system from Delphi Automotive Systems and heat spreaders in the power electronics bay supplied by dmc2 Electronic Components (DEC).

Gibbs Die Casting Corporation
369 Community Drive
Henderson KY 42420 USA

tel ++ 1 (270) 827 1801
fax ++ 1 (270) 827 7840
email
web address www.gibbsdc.com

Gibbs Die Casting Corporation has developed technology for die casting MMC feedstock. In the die casting method, the mould cavities and feed channels are evacuated; thereafter, the desired amount of material is rapidly drawn from the melt pool into the injection cylinder where it is subsequently injected into the die cavity by means of a piston and solidified under pressure. After a short dwell, the die is opened and the casting is ejected.

Gibbs Die Casting manufactured the rear brake drums for the EV-1 for General Motors using 359/SiC/10p from Alcan. Following die casting the brake drums are machined to final specification.

Golfsmith International, Inc

11000 North IH-35

Austin TX 78753 USA

tel ++ 1 (512) 837 4810 // 1 (800) 925 7709
fax ++
email customer.service@golfsmith.com
web address www.golfsmith.com

Golfsmith manufactures the Cermet 2000 driver head employing a boron carbide reinforced metal matrix composite. Boron carbide particles, 70 microns in diameter, are segregated at the surface of the head during processing to locally produce a high concentration of boron carbide. The segregation of the reinforcement to the head surface gives the club greater perimeter weighting, which apparently improves the feel and forgiveness of the club. Golfsmith markets the Cermet 2000 as benefiting from better energy transmission to the ball since the MMC is reported to absorb less energy compared to metal drivers. In addition, the MMC exhibits about 16% higher specific strength than titanium, allowing either a lower weight head or a larger head for the same weight.

Update Feb. 2003: the information above is correct as of end 2001; however, Golfsmith no longer carry the product at this time, nor any other MMC golf club components.

Goodfellow Cambridge Ltd

Ermine Business Park

Huntingdon PE29 6WR United Kingdom

tel ++	44 (0) 1480 424 800
fax ++	44 (0) 1480 424 900
email	info@goodfellow.com
web address	http://www.goodfellow.com/

Goodfellow supplies small quantities of metals, alloys, ceramics and polymers for research, development and prototyping applications. Items are available in many forms including rods, wires, tubes and foils. There is no minimum order quantity and items are in stock ready for immediate shipment worldwide. Custom-made items are available to special order, including larger quantities of standard products as well as further materials' processing.

Goodfellow stocks two aluminium matrix alloys reinforced with silicon carbide: a 2124 (aluminium-copper alloy) matrix and a 8090 (aluminium-lithium alloy) matrix that offers enhanced yield strength and lower density. Both materials are available as sheet and tube products, and the aluminium-copper matrix composite is additionally available as rod.

(updated Feb. 2003)

Goodfellow Cambridge Ltd

material (company designation):

Material 2124 / SiC / 15p

matrix Al - 4Cu - 1.5Mg - 0.7Mn

reinforcement SiC *volume percent* 15

morphology ☒ p ☐ sf ☐ cf ☐ mf ☐ w *size (μm)*

Processing

primary

specifics

secondary ☐ Casting ☐ Forging ☐ Surface Treatments ☐ Extrusion ☐ Machining ☐ Thermal Treatments

specifics

joining ☐ Brazing ☐ Mechanical ☐ Welding ☐ Other

recyclable ☐ Yes ☐ No

Properties are quoted from manufacturer as general indicators only.

Properties *isotropic* ☐ Yes ☐ No

density (g cm^{-3})	2.85	*tensile yield strength (MPa)*	400
CTE (x10^6 K^{-1})		*ultimate strength (MPa)*	610
thermal cond. (Wm^{-1}K^{-1})		*modulus (GPa)*	100
electrical cond. (%IACS)		*elongation (%)*	6

Property Advantages ☒ Density ☒ Strength ☐ Wear ☐ Fatigue ☐ Thermal Expansion ☐ Weldability ☒ Stiffness ☐ Thermal Conductivity ☐ Other

Applications

Products

Sources Goodfellow (1999).

Goodfellow Cambridge Ltd

material (company designation):

Material 8090 / SiC / 15p

matrix Al - 2.4Li - 1.4Cu - 1Mg

reinforcement SiC *volume percent* 15

morphology ☒ p ☐ sf ☐ cf ☐ mf ☐ w *size (μm)*

Processing

primary

specifics

secondary ☐ Casting ☐ Forging ☐ Surface Treatments ☐ Extrusion ☐ Machining ☐ Thermal Treatments

specifics

joining ☐ Brazing ☐ Mechanical ☐ Welding ☐ Other

recyclable ☐ Yes ☐ No

Properties are quoted from manufacturer as general indicators only.

Properties *isotropic* ☐ Yes ☐ No

			tensile
density (g cm^{-3})	2.62	*yield strength (MPa)*	500
CTE (x10^6 K^{-1})		*ultimate strength (MPa)*	550
thermal cond. (Wm^{-1}K^{-1})		*modulus (GPa)*	100
electrical cond. (%IACS)		*elongation (%)*	2.3

Property Advantages ☒ Density ☒ Strength ☐ Wear ☐ Fatigue ☐ Thermal Expansion ☐ Weldability ☒ Stiffness ☐ Thermal Conductivity ☐ Other

Applications

Products

Sources Goodfellow (1999).

Griffen Bikes

2050 East Vista Way

Vista CA 92084 USA

tel ++ 1 (805) 570 5155
fax ++
email repfree@aol.com
web address www.griffenbike.com

Griffen manufactures road racing bicycles from boron carbide reinforced 6092 aluminium. This material was developed to replace aluminium alloys in high performance applications such as seats for the (US military) Apache helicopter. Griffen produces frames and framesets in several styles: the Classic, the Vulcan and the Triton with aerodynamically-shaped down and seat tubes for improved aerodynamics.

Powders of aluminium alloy and boron carbide are blended in a jet mill to achieve a homogeneous mixture. The mixed powder is poured into a mould, degassed to remove moisture and cold isostatically pressed. The billet of compacted powder is then sintered and extruded into the various sizes of tubing necessary for construction of light, stiff and strong bicycle frames. The frames are assembled by welding, but require techniques certified by the aerospace industry. The frames are solution treated to relieve the stresses in the weld regions and subsequently artificially aged to acquire peak strength in the aluminium matrix alloy.

H.H. Racing Group			
801 N. Providence Road			
Media PA		19063	USA
tel ++	1 (610) 565 6023		
fax ++	1 (610) 565 6033		
email	dennis@hhracinggroup.com		
web address	www.hhracinggroup.com		

H.H. Racing manufactures the STD (Specific Tube Design) bicycle frame, incorporating a silicon carbide whisker-reinforced aluminium for the main triangle of the frame (top, down and seat tubes). The frame is assembled by bonding the tubes since welding negatively affects the microstructure in the weld region. This tube material cannot be exported from the US due to national security concerns, and the method of manufacture is patented. The material (2009/SiC/15w-T8) was originally marketed as SXA by Advanced Composite Materials Corporation (ACMC). ACMC developed several MMC compositions for military applications, but these materials are no longer in production, although licencing from ACMC is possible. An inventory of this material remains, allowing continued production of these frames. H.H. Racing also manufactures a seatpost from MMC.

Haute Stick Lacrosse Gear

22142 Plantino

Mission Veijo CA 92691 USA

tel ++ 1 (949) 458 5986
fax ++ 1 (949) 380 1123
email mail@hautestick.com
web address www.hautestick.com

Haute Stick Lacrosse manufactures a lacrosse shaft of Duralcan material (from Alcan). Haute Stick Lacrosse extrudes the billet material through a skull-shaped die, producing tubing with a cross-sectional geometry that improves the stiffness. The high specific stiffness of the shaft is reported to improve energy transfer from the stick to the ball during a throw.

Hi-Por Ceramics Ltd

Stubley Lane
Dronfield
Sheffield S18 1LS UK

tel ++ 44 (1246) 299 750
fax ++ 44 (1246) 299 720
email enquires@hi-por.co.uk
web address www.hi-por.com

Hi-Por Ceramics manufactures highly porous and dense ceramics. The porous ceramic materials available from Hi-Por include alumina, zirconia, silicon carbide and several others. The pores are 100% open (or interconnected) making these porous ceramics suitable for use as catalysts, filters and preforms for infiltrated MMC. Hi-Por provides these porous ceramics with densities from 10 to 95%, and average pore sizes in the range from 50 to 300 μm. The MMC produced by infiltration with aluminium are often referred to as bi-continuous or interpenetrating phase composites (IPC) because both phases are interconnected in three dimensions.

Hi-Por Ceramics is owned by the Dyson Group plc, 381 Fulwood Road, Sheffield S10 3GB, UK (www.dyson-group.com).

High Performance Materials Group (HPMG)

MSE, Inc
1300 Marrows Road
Newark DE 19711 USA

tel ++ 1 (302) 456 6200
fax ++ 1 (302) 454 1712
email kleighton@msematerials.com
web address www.lanxide.com

High Performance Materials Group, a part of MSE Inc, owns the intellectual property of the former Lanxide Corporation, in particular the PRIMEX™ and PRIMEX Cast™ processes, and the PRIMEX Concentrate™ feedstock concept.

PRIMEX™ is a pressureless metal infiltration process for the manufacture of low-cost, high-performance MMC in net or near-net shape. With this method, a ceramic preform is infiltrated with a molten aluminium alloy in a nitrogen-rich atmosphere and in the presence of magnesium (typically as an alloying addition to the melt). Complete infiltration of homogeneous preforms with reinforcement concentrations of 40 to 80 vol. % is possible. In addition, selectively reinforced components can be manufactured with the PRIMEX™ process.

The PRIMEX Cast™ process incorporates the PRIMEX™ process to produce low cost MMC materials with low to moderate reinforcement contents. The pressureless infiltration step, which achieves good wetting between the reinforcement and the melt, is followed by mechanical stirring to disperse the reinforcement in the melt to lower concentrations, typically in the range of 10 to 30 vol. %. This melt is then ready for casting by conventional methods such as gravity, die or investment casting.

The PRIMEX Concentrate™ concept allows foundries to remelt and dilute high-concentration billets for casting products with 10 to 30 vol. % ceramic. PRIMEX Concentrate™ feedstock from Lanxide was used to manufacture brake discs for the Lotus Elise (4040/SiC/30p). Lanxide and its successors have developed braking applications for motorcycles, commercial trucks and snowmobiles. Other development work includes applications of low concentration MMC for clutch plates, pistons and piston liners. dmc2 Electronic Components (DEC) markets the PRIMEX™ materials (i.e., high concentration MMC) for thermal substrate applications, for example in Motorola's Iridium satellites and in General Motors' electric vehicle (the EV-1).

High Performance Materials Group (HPMG)

material (company designation): AA4040 / SiC / 30p

Material 4040 / SiC / 30p

matrix Al-Si

reinforcement SiC *volume percent* 30

morphology ☒ p ☐ sf ☐ cf ☐ mf ☐ w *size (μm)*

Processing

primary Infiltration

specifics PRIMEX™ pressureless infiltration

secondary ☒ Casting ☐ Forging ☐ Surface Treatments ☐ Extrusion ☒ Machining ☐ Thermal Treatments

specifics PRIMEX Cast™ process

joining ☐ Brazing ☐ Mechanical ☐ Welding ☐ Other

recyclable ☐ Yes ☐ No

Properties are quoted from manufacturer as general indicators only.

Properties *isotropic* ☐ Yes ☐ No *tensile*

density (g cm^{-3})		*yield strength (MPa)*	
CTE (x10^{6} K^{-1})	14.8	*ultimate strength (MPa)*	225
thermal cond. (Wm^{-1}K^{-1})		*modulus (GPa)*	127
electrical cond. (%IACS)		*elongation (%)*	

Property Advantages ☐ Density ☐ Strength ☒ Wear ☐ Fatigue ☐ Thermal Expansion ☐ Weldability ☐ Stiffness ☐ Thermal Conductivity ☐ Other

Applications brake discs

Products Lotus Elise brake rotors

Sources Hollins, M. (1997).

High Performance Materials Group (HPMG)

material (company designation): AA7033 / Al2O3 / 25p

Material 7033 / Al2O3 / 25p

matrix Al-Mg-Zn

reinforcement Al2O3 *volume percent* 25

morphology ☒ p ☐ sf ☐ cf ☐ mf ☐ w *size (μm)*

Processing

primary Infiltration

specifics PRIMEX™ pressureless infiltration

secondary ☒ Casting ☐ Forging ☐ Surface Treatments ☐ Extrusion ☒ Machining ☐ Thermal Treatments

specifics PRIMEX Cast™ process

joining ☐ Brazing ☐ Mechanical ☐ Welding ☐ Other

recyclable ☐ Yes ☐ No

Properties are quoted from manufacturer as general indicators only.

Properties *isotropic* ☐ Yes ☐ No *tensile*

density $(g\ cm^{-3})$ *yield strength (MPa)*

CTE $(x10^{6}\ K^{-1})$ 18.0 *ultimate strength (MPa)* 180

thermal cond. $(Wm^{-1}K^{-1})$ *modulus (GPa)* 100

electrical cond. (%IACS) *elongation (%)*

Property Advantages ☐ Density ☐ Strength ☒ Wear ☐ Fatigue ☐ Thermal Expansion ☐ Weldability ☐ Stiffness ☐ Thermal Conductivity ☐ Other

Applications brake discs

Products

Sources Hollins, M. (1997).

High Performance Materials Group (HPMG)
material (company designation): AA7027 / Al2O3 / 38p

Material 7027 / Al2O3 / 38p
matrix Al-Mg-Zn
reinforcement Al2O3 *volume percent* 38
morphology ☒ p ☐ sf ☐ cf ☐ mf ☐ w *size (μm)*

Processing
primary Infiltration
specifics PRIMEX™ pressureless infiltration

secondary ☐ Casting ☐ Forging ☐ Surface Treatments ☐ Extrusion ☐ Machining ☐ Thermal Treatments
specifics PRIMEX Cast™ process

joining ☐ Brazing ☐ Mechanical ☐ Welding ☐ Other
recyclable ☐ Yes ☐ No

Properties are quoted from manufacturer as general indicators only.

Properties *isotropic* ☐ Yes ☐ No

density (g cm⁻³)		*tensile yield strength (MPa)*	
CTE ($x10^6$ K^{-1})	14.5	*ultimate strength (MPa)*	220
thermal cond. ($Wm^{-1}K^{-1}$)		*modulus (GPa)*	137
electrical cond. (%IACS)		*elongation (%)*	

Property Advantages ☐ Density ☐ Strength ☒ Wear ☐ Fatigue ☐ Thermal Expansion ☐ Weldability ☐ Stiffness ☐ Thermal Conductivity ☐ Other

Applications brake discs

Products

Sources Hollins, M. (1997).

Hitchiner Manufacturing Company, Inc

Nonferrous Division
600 Cannonball Lane
O'Fallon MO 63366 USA

tel ++ 1 (636) 272 6176
fax ++ 1 (636) 272 6180
email
web address www.hitchiner.com

Hitchiner is a world-leader in investment casting, using, in particular, its patented countergravity investment casting technology. This technology is well-suited for MMC (and a large range of conventional alloys) since it has the advantage of reducing gaseous inclusions caused by turbulent flow when filling the moulds. As the name implies, countergravity casting employs moulds suspended above the melt; the mould cavity is evacuated such that the melt can be drawn into the mould through a fill pipe due to the pressure difference established between the mould cavity and the melt. A fine grain size is achieved by heating the mould to near the solidus temperature of the melt. Excess melt flows back into the crucible after casting, an important feature of the process that minimises material losses inherent to conventional casting methods.

The Nonferrous Division at Hitchiner casts MMC components to near-net shape, sourcing particulate reinforced aluminium from both Alcan and M-Cubed. Duralcan materials are used for casting components with 20 to 30 vol. % silicon carbide and Lanxide or PRIMEX™ materials (from M-Cubed) for applications requiring 30 to 40 vol. % silicon carbide. MMC components manufactured by Hitchiner have been applied in wire bonding equipment, high-quality navigational mirrors for the military and silicon wafer processing equipment. Hitchiner's sales of countergravity cast MMC components are reportedly on the order of 2-3 million USD per year.

Honda Motor Company, Ltd (see web address)

Japan

tel ++
fax ++
email
web address www.honda.com

Honda is a pioneer in the use of MMC in the automotive industry, particularly in internal combustion engines. Although discontinued in 1985, the City Vehicle (marketed on a limited basis in Japan) employed aluminium connecting rods with continuous steel wire reinforcement. Initially, Honda developed steel fibre-reinforced aluminium components for connecting rods and driveshafts; these applications were introduced on a limited basis in Japan. In 1990, Honda launched a engine block featuring selective reinforcement of the four cylinder bores in a 2.2 litre in-line configuration; the reinforcement was a mix of Saffil® alumina fibres and graphite fibres. The success of this system led to the incorporation of the same engine into the Prelude SI model, rather than developing a new unit. In the same year, Honda released the Accord model with selectively reinforced cylinder bores. These successes were continued in 1998, with the introduction of selectively reinforced bores in the V6 engine for the Acura NSX, the first V-engine to benefit from selective reinforcement technology. More recently, Honda has launched yet another engine with selective reinforcement in the cylinder bore, the S2000 Roadster, which exhibits the highest power-per-litre output of any conventional engine in the world today.

Howmet Castings

Hot Isostatic Pressing
1600 South Warner Street
Whitehall MI 49461-1897 USA

tel ++ 1 (231) 894 7475
fax ++ 1 (231) 894 7499
email scordray@howmet.com
web address www.howmet.com

Howmet Castings, an Alcoa business, has the ability to produce MMC components and claddings by hot isostatic pressing (HIPing). Cladding the leading tip of turbine airfoils with MMC materials, for example, improves wear resistance against abrasive objects such as sand.

Inco Ltd

Suite 1500
145 King Street West
Toronto, Ontario M5H 4B7 Canada

tel ++ 1 (416) 361 7511
fax ++ 1 (416) 361 7781
email webmaster@inco.com
web address www.incosp.com

Inco Special Products, a unit of Inco Ltd, manufactures a nickel-coated graphite particle that is added to aluminium with silicon carbide to form a dual-reinforced composite called GrA-Ni®. The GrA-Ni® material is produced by a stir-casting process, and is reported to exhibit good homogeneity of the reinforcing phase in the composite due to the low density of graphite (lower than molten aluminium) preventing settling of the silicon carbide. The presence of the nickel coating results in the formation of nickel-aluminium intermetallics, which impart greater high-temperature strengthening to the material compared to conventional aluminium alloys. The presence of intermetallic also reduces the concentration of ceramic reinforcement necessary to achieve specific performance gains while improving casting and machining properties compared to composites with higher concentrations of ceramic.

GrA-Ni® displays high wear resistance due to the combination of hard SiC particles with dry lubricating graphite particles. It is intended as a direct replacement for cast iron. This material offers enhanced thermal conductivity and diffusivity in addition to lower density compared to the standard ferrous alternatives. Potential applications that have been identified by Inco include cylinder liners, brakes, valve guides, pistons and wear-resistant sleeves.

Knorr-Bremse AG

Railway Braking Systems
Moosacher Strasse 80
München 80809 Germany

tel ++ 49 (89) 35 47 0
fax ++ 49 (89) 35 47 27 67
email
web address www.knorr-bremse.com

Knorr-Bremse is aiming to introduce aluminium matrix composites into most classes of railway braking systems. The wheel-set and braking components make up over half of the total weight of bogie components; this system generates the braking force for rail transport. Knorr-Bremse has responded to interest in reducing the weight of these components by developing an aluminium-matrix composite replacement for cast iron brake discs, which satisfies the demands of high stiffness, tensile strength, thermal conductivity and wear resistance. The improved thermal conductivity of aluminium-based composites, compared to gray cast iron, allows these brake discs to be rated for greater braking forces. This results in weight savings due to a reduction in the number of discs needed (in addition to the lower density of the aluminium-based component) and in an increased load-carrying capability for the vehicle. These railway brake discs are cast by a German foundry from Duralcan material sourced from an Alcan distributor in Denmark, and machined by a subcontractor for Knorr-Bremse.

Knorr-Bremse appears to be able to justify the higher component costs of MMC discs based on long-term performance compared to conventional cast iron discs. The need for fewer discs and the energy savings associated with reduced weight translate to lower overall cost for aluminium-matrix braking components. In addition, enhanced thermal conductivity of the aluminium-based composite should contribute to longer disc life by reducing thermal stresses and hence susceptibility to cracking, particularly in intercity rail systems that experience heavy and regular braking.

Widespread commercial application of these brake discs is close to realisation, with advanced trials on freight and passenger trains in Germany and Switzerland. Urban applications are expected to expand following the successful pre-series testing of discs on the Copenhagen S-Bane (city commuter train), the first trains in the world to have aluminium-based MMC brake discs as standard equipment.

Kolbenschmidt Pierburg AG

KS Aluminium Technologie
Hafenstrasse 25
Neckarsulm 74172 Germany

tel ++ 49 (7132) 33 0
fax ++ 49 (7132) 33 2722
email info-alucasting@kolbenschmidt-pierburg.com
web address www.kolbenschmidt.de

KS Aluminium Technologie is a division of the German engineering firm Kolbenschmidt Pierburg and is one of the leaders in employing MMC technology for cylinder bore surfaces. KS Aluminium Technologie has developed the LOKASIL® technology for selective reinforcement of cylinder bores for high performance passenger automobiles: LOKASIL I® contains 5 vol. % alumina and 15 vol. % silicon; LOKASIL II® has only 25 vol. % silicon. Preforms of LOKASIL I® are prepared from a mixture of water, alumina fibres, silicon particulate, and a binder that are vacuum pumped onto a sieve, resulting in the fibres depositing primarily tangentially on the sieve circumference. Preforms of LOKASIL II® are manufactured by a freeze casting process that produces a porous structure. Both techniques produce a green preform that requires sintering prior to infiltration. The engine blocks are produced by squeeze casting with alloy 226 (Al-9Si-3Cu) concomitant with infiltration of the net-shaped LOKASIL® preforms at the cylinder bore.

The LOKASIL® technology replaces engine blocks cast from 390 alloy (Al-17Si-4Cu-Mg), and makes gray cast iron inserts unnecessary. This also avoids potential problems associated with Young's modulus mismatch and thermal expansion mismatch between the gray cast iron cylinder liners and aluminium engine block. The LOKASIL® technology has been proven in the 2.5-litre engine for the Porsche Boxster, and more recently in the 3.4-litre engine for the 911 Carrera. In these applications the LOKASIL II® system was chosen because thicker, lower strength preforms could be used for the cylinder arrangement; LOKASIL I® is a more space-efficient, higher-strength, solution.

In addition to the LOKASIL® technology, Kolbenschmidt Pierburg casts the rear brake drums for Volkswagen's Lupo 3-litre model. These drums are die cast using A359/SiC/20p from Alcan (Duralcan material). Kolbenschmidt Pierburg also produces selectively reinforced pistons for diesel engines.

Kolbenschmidt Pierburg AG

material (company designation): LOKASIL® I

Material	Al / Al2O3+ Si / 5sf+15p		
matrix	Al - 9Si - 3Cu		
reinforcement	Al2O3 and Si	*volume percent*	5sf+15
morphology	☒ p ☐ sf ☐ cf ☐ mf ☐ w	*size (μm)*	Si: 30 -

Processing	
primary	Infiltration
specifics	selectively-reinforced squeeze casting
secondary	☐ Casting ☐ Forging ☐ Surface Treatments ☐ Extrusion ☐ Machining ☐ Thermal Treatments
specifics	
joining	☐ Brazing ☐ Mechanical ☐ Welding ☐ Other
recyclable	☒ Yes ☐ No

Properties are quoted from manufacturer as general indicators only.

Properties	*isotropic* ☐ Yes ☐ No		*tensile*
density (g cm⁻³)	2.71	*yield strength (MPa)*	140-170
CTE ($x10^6$ K^{-1})	17-18 (20-200°C)	*ultimate strength (MPa)*	180-230
thermal cond. ($Wm^{-1}K^{-1}$)		*modulus (GPa)*	85-95
electrical cond. (%IACS)		*elongation (%)*	0.1-0.9
Property Advantages	☐ Density ☐ Strength ☒ Wear ☒ Fatigue ☒ Thermal Expansion ☐ Weldability ☐ Stiffness ☐ Thermal Conductivity ☐ Other		

Applications	selective reinforcement of cylinder bores for automobiles
Products	
Sources	

Kolbenschmidt Pierburg AG

material (company designation): LOKASIL® II

Material Al / Si / 25p

matrix Al - 9Si - 3Cu

reinforcement Si *volume percent* 25

morphology ☒ p ☐ sf ☐ cf ☐ mf ☐ w *size (μm)* 30-70

Processing

primary Infiltration

specifics selectively-reinforced squeeze casting

secondary ☐ Casting ☐ Forging ☐ Surface Treatments ☐ Extrusion ☐ Machining ☐ Thermal Treatments

specifics

joining ☐ Brazing ☐ Mechanical ☐ Welding ☐ Other

recyclable ☒ Yes ☐ No

Properties are quoted from manufacturer as general indicators only.

Properties *isotropic* ☐ Yes ☐ No

			tensile
density (g cm^{-3})	2.65	*yield strength (MPa)*	
CTE (x10^{6} K^{-1})	17.5 (20-100°C)	*ultimate strength (MPa)*	140-160
thermal cond. (Wm^{-1}K^{-1})		*modulus (GPa)*	75-85
electrical cond. (%IACS)		*elongation (%)*	0.06-0.12

Property Advantages ☐ Density ☐ Strength ☒ Wear ☒ Fatigue ☒ Thermal Expansion ☐ Weldability ☐ Stiffness ☐ Thermal Conductivity ☐ Other

Applications selective reinforcement of cylinder bores for automobiles

Products		
	Boxster	cylinder bores
	911 Carreras	cylinder bores

Sources KS Aluminium Technologie AG.

Kubota Corporation

1-3 Nihonbashi muromachi 3-chrome
Chuo-Ku
Tokyo 103-8310 Japan

tel ++ 81 (3) 3245 3568
fax ++ 81 (3) 3245 3565
email sinsozai@kukoba.co.jp
web address www.kubota.co.jp/infra/shinsozai/ceramics/mcslee

The New Consolidated Materials Department at Kubota has developed both materials and process technologies related to the manufacture of MMC materials by powder metallurgical methods, in particular hot and cold isostatic pressing techniques (HIPing and CIPing). Kubota produces MMC materials based on titanium and cobalt alloy matrices reinforced with ceramic particulate. The reinforcement can be varied to meet the needs of the application and include TiC, SiC, $MoSi_2$ and TiB.

The materials and components produced by Kubota are designed to serve the aluminium die casting industry. In shot-sleeves for high-pressure die casting, the low thermal conductivity of the titanium-matrix composition maintains the temperature of the melt inside the sleeve during casting. The material also exhibits excellent corrosion resistance, thus improving the lifetime of the sleeve and the quality of the casting.

US offices:

Kubota Corp.
320 Park Avenue
23rd Floor
New York NY 10022
tel: 1 (212) 355 2440
fax: 1 (212) 355 5887

Offices and divisions in Canada, UK and Germany; consult website for further details.

Kubota Corporation

material (company designation): MC-302

Material Co / / p

matrix unspecified grade Co

reinforcement *volume percent*

morphology ☒ p ☐ sf ☐ cf ☐ mf ☐ w *size (μm)*

Processing

primary Powder Metallurgy

specifics Hot Isostatic Pressing (HIPing)

secondary ☐ Casting ☐ Forging ☐ Surface Treatments ☐ Extrusion ☐ Machining ☐ Thermal Treatments

specifics

joining ☐ Brazing ☐ Mechanical ☐ Welding ☐ Other

recyclable ☐ Yes ☐ No

Properties are quoted from manufacturer as general indicators only.

Properties *isotropic* ☐ Yes ☐ No *tensile*

density (g cm⁻³) 8.4 *yield strength (MPa)*

CTE (x10⁻⁶ K⁻¹) 12.8 *ultimate strength (MPa)*

thermal cond. (Wm⁻¹K⁻¹) 13 *modulus (GPa)* 260

electrical cond. (%IACS) *elongation (%)*

Property Advantages ☐ Density ☐ Strength ☒ Wear ☐ Fatigue ☐ Thermal Expansion ☐ Weldability ☐ Stiffness ☐ Thermal Conductivity ☐ Other

Applications sleeves for aluminium die casting

Products

Sources

Kubota Corporation

material (company designation): MC-230

Material Ti / / p

matrix unspecified grade Ti

reinforcement *volume percent*

morphology ☒ p ☐ sf ☐ cf ☐ mf ☐ w *size (μm)*

Processing

primary Powder Metallurgy

specifics Hot Isostatic Pressing (HIPing)

secondary ☐ Casting ☐ Forging ☐ Surface Treatments ☐ Extrusion ☐ Machining ☐ Thermal Treatments

specifics

joining ☐ Brazing ☐ Mechanical ☐ Welding ☐ Other

recyclable ☐ Yes ☐ No

Properties are quoted from manufacturer as general indicators only.

Properties *isotropic* ☐ Yes ☐ No *tensile*

density ($g\ cm^{-3}$) 5.2 *yield strength (MPa)*

CTE ($x10^6\ K^{-1}$) 7.5 *ultimate strength (MPa)*

thermal cond. ($Wm^{-1}K^{-1}$) 13 *modulus (GPa)* 130

electrical cond. (%IACS) *elongation (%)*

Property Advantages ☐ Density ☐ Strength ☒ Wear ☐ Fatigue ☐ Thermal Expansion ☐ Weldability ☐ Stiffness ☐ Thermal Conductivity ☐ Other

Applications sleeves for aluminium die casting

Products

Sources

Kubota Corporation

material (company designation): MC-210

Material Ti / / p

matrix unspecified grade Ti

reinforcement *volume percent*

morphology ☒ p ☐ sf ☐ cf ☐ mf ☐ w *size (μm)*

Processing

primary Powder Metallurgy

specifics Hot Isostatic Pressing (HIPing)

secondary ☐ Casting ☐ Forging ☐ Surface Treatments ☐ Extrusion ☐ Machining ☐ Thermal Treatments

specifics

joining ☐ Brazing ☐ Mechanical ☐ Welding ☐ Other

recyclable ☐ Yes ☐ No

Properties are quoted from manufacturer as general indicators only.

Properties *isotropic* ☐ Yes ☐ No *tensile*

density (g cm^{-3}) 5.1 *yield strength (MPa)*

CTE (x10^{6} K^{-1}) 7.5 *ultimate strength (MPa)*

thermal cond. (Wm^{-1}K^{-1}) 9 *modulus (GPa)* 130

electrical cond. (%IACS) *elongation (%)*

Property Advantages ☐ Density ☐ Strength ☒ Wear ☐ Fatigue ☐ Thermal Expansion ☐ Weldability ☐ Stiffness ☐ Thermal Conductivity ☒ Other

Applications sleeves for aluminium die casting

Products

Sources

Laser Machining, Inc

500 Laser Drive

Somerset WI 54025 USA

tel ++ 1 (715) 247 3285
fax ++ 1 (715) 247 5650
email DPlourde@lasermachining.com
web address www.lasermachining.com

Laser Machining provides both process capability and equipment for laser processing of materials. Laser cutting technology has been proven for machining particle-reinforced aluminium, and it can provide a cost-effective method compared to tool machining of MMC. The benefits of laser machining compared to conventional techniques include the ability to cut very hard or abrasive materials while always maintaining a sharp "cutting edge", and cutting at high speeds limited primarily by the power available from the laser.

Lockheed Martin Aeronautics Company
Lockheed Boulevard
Fort Worth TX 76108 USA

tel ++ 1 (817) 777 2000
fax ++
email
web address www.lockheedmartin.com

Lockheed Martin is a major aerospace company serving both civilian and governmental customers. Through the Title III program sponsored by the US government, Lockheed Martin has contributed to the development and qualification of particle-reinforced aluminium (also called discontinuously reinforced aluminium or DRA) for aerospace applications, in particular for the F-16 Falcon fighter aircraft.

Ventral fins provide lateral stability on fighter aircraft during high angle attack manoeuvres and on the F-16 these were initially constructed of aluminium sheets bonded to a honeycomb core. These fins suffer severe in-flight buffeting that ultimately has caused failure after 400 hours in service due to insufficient stiffness. Particle-reinforced aluminium sheets have stiffness of up to 40% more than the unreinforced aluminium alloy, although the densities are similar. Replacing the aluminium sheets with MMC and improving the leading edge of the component contributes to lifetimes of over 6000 hours for the ventral fins. This performance enhancement is projected to save the US Air Force 21 million USD in life cycle costs.

MMC sheet is used elsewhere on the F-16. With the implementation of improved door fasteners, fuel access cover doors are required to bear load and reduce the aircraft skin stress. This design change led to overloading and cracking of the doors. Doubling the thickness of the cover doors would have solved this problem but with an unacceptable weight penalty. Direct replacement of the alloy doors with MMC doors, led to sufficient strengthening of the component with no additional weight and a reduction in skin stress by at least 38%.

The ammunition rack for the AC-130 gunship is another successful application of MMC on Lockheed Martin aircraft. The use of PRM reduced the weight of the rack by 400 kg in addition to providing greater damage tolerance in the component.

London & Scandinavian Metallurgical Co Ltd (LSM)

Fullerton Road
Rotherham
South Yorkshire S60 1DL UK

tel ++ 44 (1709) 828 500
fax ++ 44 (1709) 828 367
email sales@lsm.co.uk
web address www.lsm.co.uk

London and Scandinavian Metallurgical has utilised its knowledge of grain refiners for aluminium alloys to develop a range of aluminium-based MMC materials that incorporate titanium diboride (TiB_2) ceramic particles as the reinforcing phase. The reinforcing phase is formed in-situ within the melt by introducing molten salts (K_2TiF_6 and KBF_4) into an aluminium melt where they react to form the boride. The resultant microstructure exhibits a homogenous dispersion of reinforcing particles, and the interfacial bonding between matrix and reinforcement is good. LSM offers several compositions as waffle plate:

(i) Al / TiB2 / 10p
(ii) Al-12Si / TiB2 / 6p
(iii) Al-4.5Cu / TiB2 / 8p

The material can be remelted, formed, machined and welded by conventional means. The fine size (2-5 μm) and low concentration of the reinforcement improve the machinability of these materials compared to other common MMC.

M-Cubed Technologies, Inc
921 Main Street
Monroe CT 06468 USA

tel ++ 1 (203) 452 2333
fax ++ 1 (203) 452 2335
email info@mmmt.com
web address www.mmmt.com

M-Cubed evolved from the bankruptcy of Lanxide Corporation and uses Lanxide technology, in particular the PRIMEX™ pressureless infiltration process. M-Cubed has also developed other technologies, and produces various advanced materials systems such reaction-bonded composites.

M-Cubed targets a diverse range of applications within automotive, aerospace, electronics, and military industries. Aluminium, reinforced with high concentration of silicon carbide, as produced by the PRIMEX™ process, are suited for applications that require high dimensional stability such as silicon wafer chucks, structures for optical systems and precision motion control, as well as for thermal management. M-Cubed also uses the PRIMEX Cast™ process for production of components with lower fractions of reinforcement and supplies material to Hitchiner for countergravity investment casting.

M-Cubed Technologies, Inc
material (company designation): ASC-701

Material Al / SiC / 70p

matrix unspecified grade Al
reinforcement SiC *volume percent* 70
morphology ☒ p ☐ sf ☐ cf ☐ mf ☐ w *size (μm)*

Processing
primary Infiltration
specifics PRIMEX™ pressureless infiltration

secondary ☐ Casting ☐ Forging ☐ Surface Treatments ☐ Extrusion ☐ Machining ☐ Thermal Treatments
specifics

joining ☐ Brazing ☐ Mechanical ☐ Welding ☐ Other
recyclable ☐ Yes ☐ No

Properties are quoted from manufacturer as general indicators only.

Properties *isotropic* ☐ Yes ☐ No *tensile*

density (g cm^{-3})	3.01	*yield strength (MPa)*	
CTE (x10^{6} K^{-1})	6.2	*ultimate strength (MPa)*	
thermal cond. (Wm^{-1}K^{-1})	170	*modulus (GPa)*	270
electrical cond. (%IACS)		*elongation (%)*	0.1

Property Advantages ☒ Density ☐ Strength ☐ Wear ☐ Fatigue ☒ Thermal Expansion ☐ Weldability ☐ Stiffness ☒ Thermal Conductivity ☐ Other

Applications thermal management in electronics

Products

Sources

Mahle GmbH

Pragstrasse 26-46

Stuttgart 70376 Germany

tel ++ 49 (711) 501 0
fax ++ 49 (711) 501 20 07
email ubl.kontakt@mahle.com
web address www.mahle.com

Mahle manufactures fibre-reinforced aluminium pistons for truck engines by a gravity casting technique. The pistons are selectively reinforced at the crown with either alumina or silicon carbide short fibres.

Mahle also produces wear-resistant coatings to improve engine performance: Nikasil® coatings employ an electrodeposited nickel reinforced with silicon carbide particles. The goal of this technology is to increase the operating temperature of engine components without excessive wear and with reduced emissions.

Mark Williams Enterprises

765 South Pierce Avenue

Louisville CO 80027 USA

tel ++	1 (303) 665 6901 // 1 (800) 525 1963
fax ++	1 (303) 665 7021
email	info@markwilliams.com
web address	www.markwilliams.com

Mark Williams manufactures driveshafts for motor racing, in particular drag racing. Their products include a metal matrix composite driveshaft, made using Duralcan material from Alcan. The particle-reinforced aluminium from Alcan is extruded, then fitted with aluminium weld yokes that have been manufactured by Mark Williams and MIG-welded in place. The resulting driveshaft has higher specific strength and stiffness than aluminium or steel shafts, allowing for a reduction in wall thickness and weight of the assembly. The enhanced specific properties offer the additional benefit that the critical rotation speed at which the shaft exhibits harmonic vibrations is increased, thus reducing the potential for structure instability. The driveshafts are available in diameters of 3.56 and 4 inches; the latter dimension is required to achieve the necessary stiffness when the length of the driveshaft is greater than 50 inches.

Materiali Metallici Compositi srl (MMC)

Via Grado 76

Monfalcone GO 34074 Italy

tel ++ 39 (481) 48 23 40
fax ++
email info@mmcitaly.it
web address www.mmcitaly.it

Material Metallici Compositi (MMC) has obtained the exclusive rights to use Liquid Pressure Forming (LPF), as designed by Composite Metal Technology (CMT), to manufacture metal matrix composites. The LPF machines are manufactured by MIR SpA.

Mavic SA

Les Croisclets
Metz-Tessy
Annecy Cedex 9 74996 France

tel ++ 33 450657171
fax ++ 33 450657172
email
web address www.mavic.com

Mavic specialises in wheel technology for bicycles, and it has developed an MMC coating for bicycle rims. The Ceramic 2 coating is produced by plasma-spraying titanium and alumima reinforcing particles onto the sidewall of the rims to form the braking surface. The reinforcement acts to dramatically increase the hardness of the rim compared to conventional aluminium alloy, offering greater wear resistance and extended rim life. In addition to these advantages, the Ceramic 2 coating improves the resistance of the rim to surface and shock damage, and reduces braking distances in both wet and dry conditions.

Mavic SA

material (company designation): Ceramic 2

Material Ti / Al2O3 / p

matrix unspecified grade Al

reinforcement Al2O3 *volume percent*

morphology ☒ p ☐ sf ☐ cf ☐ mf ☐ w *size (μm)*

Processing

primary Plasma Spray

specifics plasma beam torch spreading of titanium and alumina

secondary ☐ Casting ☐ Forging ☐ Surface Treatments ☐ Extrusion ☐ Machining ☐ Thermal Treatments

specifics

joining ☐ Brazing ☐ Mechanical ☐ Welding ☐ Other

recyclable ☐ Yes ☐ No

Properties are quoted from manufacturer as general indicators only.

Properties *isotropic* ☐ Yes ☐ No *tensile*

density (g cm^{-3}) *yield strength (MPa)*

CTE (x10^{6} K^{-1}) *ultimate strength (MPa)*

thermal cond. (Wm^{-1}K^{-1}) *modulus (GPa)*

electrical cond. (%IACS) *elongation (%)*

Property Advantages ☐ Density ☐ Strength ☒ Wear ☐ Fatigue ☐ Thermal Expansion ☐ Weldability ☐ Stiffness ☒ Thermal Conductivity ☐ Other

Applications coating for sidewalls of bicycles rims

Products bicycle rims with coating on sidewalls

Sources

Mazda Motor Corporation

Japan

tel ++
fax ++
email
web address

Following the lead of the two major Japanese auto manufacturers, Mazda has developed its own manufacturing capability for the production of particle-reinforced aluminium pistons for direct injection diesel engines. There are also plans to utilise the technology for other parts requiring enhanced strength and wear resistance.

The Air-Assisted Infiltration process developed at Mazda makes it possible to infiltrate preforms of porous metal or ceramic particles with 10 atmospheres of pressure. The technology reportedly reduces the manufacturing costs of MMC components compared to conventional squeeze casting techniques, in part because the process can be adapted to existing systems for general purpose casting. Increased productivity also results from simpler mould design and casting of multiple parts in one cycle.

Metal Matrix Cast Composites (MMCC)
101 Clematis Avenue
Waltham MA 02453-7012 USA

tel ++ 1 (781) 893 4449
fax ++ 1 (781) 893 7230
email jcornie@mmccinc.com
web address www.mmccinc.com

Metal Matrix Cast Composites (MMCC) manufactures specialised components, primarily for the US government, from a wide range of MMC material systems. Additionally, it offers design, prototyping and manufacturing services to the MMC industry in several industrial sectors. The patented Advanced Pressure Infiltration Casting method of MMCC is the cornerstone of its activities; this net shape process employs 3Dprinting™ technology to produce preforms with intricate geometrical shapes for subsequent infiltration.

MMCC employs a variety of MMC material systems, including matrices of both aluminium and copper, and reinforcements of both particulate (silicon carbide, alumina and boron carbide) and discontinuous carbon fibres for thermal management applications. MMCC has attempted to capitalise on the potential for weight reduction intrinsic to MMC components compared to ferrous solutions, particularly in moving parts; compressor rotors, brake callipers and connecting rods have all been produced from aluminium matrix composite (2214/Al2O3/55p). Annual sales at MMCC are reportedly between 2.5-5 million dollars; the distribution between governmental and commercial sales is not given.

Metalba SpA

Viale Vicenza 71

Bassano del Grappa VI 36061 Italy

tel ++	39 (424) 25 23 00
fax ++	39 (424) 50 39 44
email	info@metalba.it
web address	www.metalba.it

Metalba serves industries requiring extruded and drawn aluminium alloy products, including particle reinforced aluminium primarily from Alcan.

Metallic Composites for the 21st Century, Inc (MC-21)
5100 Convair Drive

Carson City NV 89706-0425 USA

tel ++ 1 (775) 841 7115
fax ++ 1 (775) 841 7117
email MC21dms@worldnet.att.net
web address http://www.mc21inc.com

MC-21 manufactures particle-reinforced aluminium by stir-casting. The founders, D.Schuster and M.Skibo, have considerable experience in the MMC industry as they developed the stir-casting process currently used by Alcan.

MC-21 has developed a novel roll casting process to produce composite sheet suitable for thermal management components for the integrated circuits industry. Sheet material is cast as 25 mm thick slabs by a vacuum-assisted stirring process that ensures good wetting and bonding between the aluminium matrix and the reinforcement (SiC or Al_2O_3). The cast material is subsequently hot-rolled to a thickness of 0.5 mm. MC-21 is capable of producing batches of 60-600 kg with SiC concentrations up to 45 vol. %; (50 wt%) SiC; however, sheet reinforced by 20 or 30 vol. % particles are standard products for subsequent stamping operations. These MMC sheet materials have thermophysical properties (thermal expansion and conductivity) suitable for thermal spreaders and lids.

MC-21 has also utilised its stir casting technology to manufacture low-cost automotive components such as brake discs. By using lower cost SiC powder with a wider particle distribution, MC-21 claims that Al/SiC/20p can be produced for less than 2.40 USD per kg. Preliminary foundry studies have proved successful; the foundry that manufactures the rear brake discs for the Plymouth Prowler (Duralcan material) had positive results casting brake discs from 227 kg of ingot from MC-21. The stir-casting technology is modular and can be integrated into standard foundry operations; thus composite ingots do not need to be purchased and remelted, improving productivity by making the process semi-continuous. The modular approach also allows foundries to scale this MMC technology to their needs, reducing inventory and overhead costs. This technology is available under licence from MC-21.

Updated Feb. 2003; new information since Nov. 2001 is vol. % SiC up to 45% (formerly 35%), the ability to roll Al-SiC sheet up to 30 vol. % SiC (formerly 20%), and the website.

Metallic Composites for the 21st Century, Inc (MC-21) material (company designation):

Material Al / SiC / 20p / sheet

matrix unspecified grade Al

reinforcement SiC *volume percent* 20

morphology ☒ p ☐ sf ☐ cf ☐ mf ☐ w *size (μm)*

Processing

primary Stir Casting

specifics vacuum-assisted stirring

secondary ☐ Casting ☐ Forging ☐ Surface Treatments ☐ Extrusion ☐ Machining ☐ Thermal Treatments

specifics cast 25mm thick slabs and roll to final thickness from

joining ☐ Brazing ☐ Mechanical ☐ Welding ☐ Other

recyclable ☐ Yes ☐ No

Properties are quoted from manufacturer as general indicators only.

Properties *isotropic* ☒ Yes ☐ No *tensile*

density (g cm^{-3}) *yield strength (MPa)*

CTE (x10^{6} K^{-1}) 15 *ultimate strength (MPa)*

thermal cond. (Wm^{-1}K^{-1}) 185 *modulus (GPa)* 113.8

electrical cond. (%IACS) *elongation (%)*

Property Advantages ☐ Density ☐ Strength ☐ Wear ☐ Fatigue ☒ Thermal Expansion ☐ Weldability ☐ Stiffness ☒ Thermal Conductivity ☐ Other

Applications thermal spreaders, heat sinks, microelectronic package stiffeners, bicycle disk brake rotors, loudspeaker cones

Products

Sources

Metallic Composites for the 21st Century, Inc (MC-21) material (company designation):

Material Al / SiC / 30p / sheet

matrix unspecified grade Al

reinforcement SiC *volume percent* 30

morphology ☒ p ☐ sf ☐ cf ☐ mf ☐ w *size (μm)*

Processing

primary Stir Casting

specifics vacuum-assisted stirring

secondary ☐ Casting ☐ Forging ☐ Surface Treatments ☐ Extrusion ☐ Machining ☐ Thermal Treatments

specifics cast 25mm thick slabs and roll to final thickness from

joining ☐ Brazing ☐ Mechanical ☐ Welding ☐ Other

recyclable ☐ Yes ☐ No

Properties are quoted from manufacturer as general indicators only.

Properties *isotropic* ☒ Yes ☐ No *tensile*

density (g cm⁻³) — written as *density* ($g\ cm^{-3}$) *yield strength (MPa)*

CTE ($x10^6\ K^{-1}$) 13-14 *ultimate strength (MPa)*

thermal cond. ($Wm^{-1}K^{-1}$) 180-185 *modulus (GPa)* 138

electrical cond. (%IACS) *elongation (%)*

Property Advantages ☐ Density ☐ Fatigue ☐ Stiffness ☐ Strength ☒ Thermal Expansion ☒ Thermal Conductivity ☐ Wear ☐ Weldability ☐ Other

Applications thermal spreaders, baseplates, heat sinks, microelectronic package stiffeners, robotic wafer transfer blades

Products

Sources

Metallic Composites for the 21st Century, Inc (MC-21)
material (company designation):

Material Al / SiC / 10-40p / ingot or cast component

matrix unspecified grade Al
reinforcement SiC *volume percent* 10-40
morphology ☒ p ☐ sf ☐ cf ☐ mf ☐ w *size (μm)*

Processing
primary Stir Casting
specifics vacuum-assisted stirring

secondary ☒ Casting ☐ Forging ☐ Surface Treatments ☐ Extrusion ☐ Machining ☐ Thermal Treatments
specifics Remelt ingots are available at 10, 20, 30 or 40 % SiC;

joining ☒ Brazing ☐ Mechanical ☐ Welding ☒ Other
recyclable ☐ Yes ☐ No

Properties are quoted from manufacturer as general indicators only.

Properties *isotropic* ☒ Yes ☐ No *tensile*
density (g cm^{-3}) 2.7 - 2.9 *yield strength (MPa)*
CTE (x10^6 K^{-1}) 11-16 *ultimate strength (MPa)*
thermal cond. (Wm^{-1}K^{-1}) 180-200 *modulus (GPa)* 86- 165
electrical cond. (%IACS) *elongation (%)*

Property Advantages ☐ Density ☐ Strength ☐ Wear ☐ Fatigue ☒ Thermal Expansion ☐ Weldability ☐ Stiffness ☒ Thermal Conductivity ☐ Other

Applications thermal spreaders, heat sinks, microelectronic package stiffeners, loudspeaker frames, automotive disk brake rotors, applications where stiffness higher than Al and/or CTE lower than Al are required.

Products

Sources

Metso Powdermet Oy

PO Box 306
Lokomonkatu 3
Tampare 33101 Finland

tel ++ 358 20 484 120
fax ++ 358 20 484 4707
email
web address www.powdermet.com

Metso Powdermet is active in powder metallurgical processing of materials, including steel-matrix composites. Metso specialises in hot isostatic pressing (HIPing) and operate in a strategical alliance with Bodycote HIP, therefore allowing Metso materials to be manufactured on a large scale throughout Europe and the United States.

Metso offers two distinct families of MMC materials aimed at applications where wear and corrosion resistance are important: a duplex stainless steel reinforced with chromium carbide particulate called Ralloy® DMMC, and an austenitic stainless steel reinforced with alumina particulate called Ralloy® AMMC. These MMC materials are produced in bulk (requiring secondary processing), or alternatively as cladding to impart wear and corrosion resistance to other engineering materials. Bulk processing places manufacturing constraints on application of these materials for two reasons: (i) these materials are difficult to machine, requiring specialised and often expensive techniques, and (ii) these materials cannot be welded to steel or stainless steel, hence requiring HIP to join these MMC to other components. A raiser tube for the petrochemical industry, for example, has been clad on the inside surface with austenitic stainless steel reinforced with 20 vol. % alumina (Ralloy® AMMC: AISI 316/Al2O3/20p) for an application where erosion was prevalent under intense turbulence.

Metso Powdermet Oy

material (company designation): Ralloy® DMMC 20

Material Duplok 22 / Cr3C2 / 20p

matrix Fe

reinforcement Cr3C2 *volume percent* 20

morphology ☒ p ☐ sf ☐ cf ☐ mf ☐ w *size (μm)*

Processing

primary Powder Metallurgy

specifics Hot Isostatic Pressing (HIPing)

secondary ☐ Casting ☐ Forging ☐ Surface Treatments ☐ Extrusion ☐ Machining ☐ Thermal Treatments

specifics

joining ☐ Brazing ☐ Mechanical ☐ Welding ☐ Other

recyclable ☐ Yes ☐ No

Properties are quoted from manufacturer as general indicators only.

Properties *isotropic* ☐ Yes ☐ No

density (g cm^{-3})	8.05		*tensile*
CTE (x10^{6} K^{-1})	12.1 (20-400°C)	*yield strength (MPa)*	
thermal cond. (Wm^{-1}K^{-1})		*ultimate strength (MPa)*	500-600
electrical cond. (%IACS)		*modulus (GPa)*	203
		elongation (%)	

Property Advantages ☐ Density ☐ Fatigue ☐ Stiffness ☐ Strength ☐ Thermal Expansion ☐ Thermal Conductivity ☒ Wear ☐ Weldability ☒ Other

Applications wear-resistant cladding

Products

Sources

Metso Powdermet Oy

material (company designation): Ralloy® DMMC 30

Material Duplok 22 / Cr3C2 / 30p

matrix Fe

reinforcement Cr3C2 *volume percent* 30

morphology ☒ p ☐ sf ☐ cf ☐ mf ☐ w *size (μm)*

Processing

primary Powder Metallurgy

specifics Hot Isostatic Pressing (HIPing)

secondary ☐ Casting ☐ Forging ☐ Surface Treatments ☐ Extrusion ☐ Machining ☐ Thermal Treatments

specifics

joining ☐ Brazing ☐ Mechanical ☐ Welding ☐ Other

recyclable ☐ Yes ☐ No

Properties are quoted from manufacturer as general indicators only.

Properties *isotropic* ☐ Yes ☐ No

			tensile
density ($g\ cm^{-3}$)	8.05	*yield strength (MPa)*	
CTE ($x10^6\ K^{-1}$)	12.1 (20-400°C)	*ultimate strength (MPa)*	500-600
thermal cond. ($Wm^{-1}K^{-1}$)		*modulus (GPa)*	203
electrical cond. (%IACS)		*elongation (%)*	

Property Advantages ☐ Density ☐ Strength ☒ Wear ☐ Fatigue ☐ Thermal Expansion ☐ Weldability ☐ Stiffness ☐ Thermal Conductivity ☒ Other

Applications wear resistant cladding

Products

Sources

Metso Powdermet Oy

material (company designation): Ralloy® AMMC 30

Material 316 / Al2O3 / 30p

matrix Fe: AISI -specified stainless steel

reinforcement Al2O3 *volume percent* 30

morphology ☒ p ☐ sf ☐ cf ☐ mf ☐ w *size (μm)*

Processing

primary Powder Metallurgy

specifics Hot Isostatic Pressing (HIPing)

secondary ☐ Casting ☐ Forging ☐ Surface Treatments ☐ Extrusion ☐ Machining ☐ Thermal Treatments

specifics

joining ☐ Brazing ☐ Mechanical ☐ Welding ☐ Other

recyclable ☐ Yes ☐ No

Properties are quoted from manufacturer as general indicators only.

Properties *isotropic* ☐ Yes ☐ No *tensile*

density (g cm^{-3})	7.2	*yield strength (MPa)*	
CTE (x10^{6} K^{-1})	13.8 (20-400°C)	*ultimate strength (MPa)*	250-300
thermal cond. (Wm^{-1}K^{-1})		*modulus (GPa)*	203
electrical cond. (%IACS)		*elongation (%)*	

Property Advantages ☐ Density ☐ Strength ☒ Wear ☐ Fatigue ☐ Thermal Expansion ☐ Weldability ☐ Stiffness ☐ Thermal Conductivity ☒ Other

Applications cladding for erosion resistance, wear parts in fans, mills, cutters, erosion shields in boilers, pump and valve components

Products

Sources

Millennium Materials, Inc (MM)

642 Barbrow Lane

Knoxville TN 37932 USA

tel ++ 1 (865) 966 2170
fax ++ 1 (865) 966 7477
email samcweaver@aol.com
web address www.millenmat.com

Millennium Materials (MM) manufactures advanced materials, including ceramic and metal matrix composites, as well as ceramic particulate and whisker.

MM produces a family of MMC by stir-casting called Carolite®: aluminium matrix (specifically A356) and silicon hexaboride (SiB_6) reinforcing particulate. The concentration of the SiB_6 phase (trade named Hexon®) can be optimised for specific applications, reportedly in the range of 5 to 40 vol. %. Silicon hexaboride has a specific gravity similar to that of molten aluminium alloys, thus the reinforcement remains in suspension with minimum agitation, facilitating casting by conventional techniques and maintaining uniform distributions of reinforcement in the MMC. Since the reinforcement does not segregate, Carolite® can be recycled. In addition, the material is not adverse to fusion welding.

Carolite® (A356/ SiB6/40p) is being developed as a substrate material for computer hard disks, due to its high specific elastic modulus and good damping properties. These properties are essential to this application because they enable faster spin-up and spin-down times, and more stable rotation with less disc flutter, hence providing lower fly heights, lower seek times, and higher memory density.

MM also offers ceramic preforms for infiltration in a range of densities and shapes that can be optimised for specific requirements. The silicon carbide composite preforms are sufficiently robust to withstand automated handling and high-pressure infiltration without breakage. Other ceramics reinforcements are also available: silicon carbide platelets and fibres, and boron carbide whiskers and platelets.

MM is owned by the Dyson Group plc, 381 Fulwood Road, Sheffield S10 3GB, UK (www.dyson-group.com).

Millennium Materials, Inc (MM)

material (company designation): Carolite®

Material A356 / SiB6 / 10p

matrix Al alloy

reinforcement SiB6 *volume percent* 10

morphology ☒ p ☐ sf ☐ cf ☐ mf ☐ w *size (μm)*

Processing

primary Stir Casting

specifics

secondary ☐ Casting ☐ Forging ☐ Surface Treatments ☐ Extrusion ☐ Machining ☐ Thermal Treatments

specifics

joining ☐ Brazing ☐ Mechanical ☒ Welding ☐ Other

recyclable ☒ Yes ☐ No

Properties are quoted from manufacturer as general indicators only.

Properties *isotropic* ☐ Yes ☐ No

		tensile	
density (g cm^{-3})		*yield strength (MPa)*	
CTE (x10^6 K^{-1})	15.5	*ultimate strength (MPa)*	259
thermal cond. (Wm^{-1}K^{-1})	146	*modulus (GPa)*	
electrical cond. (%IACS)		*elongation (%)*	

Property Advantages ☒ Density ☐ Fatigue ☒ Stiffness ☒ Strength ☒ Thermal Expansion ☒ Thermal Conductivity ☒ Wear ☒ Weldability ☐ Other

Applications sailboat masts and other lightweight structural components

Products

Sources Millennium Materials, Inc (2000).

Millennium Materials, Inc (MM)

material (company designation): Carolite®

Material A356 / SiB6 / 15p

matrix Al alloy

reinforcement SiB6 *volume percent* 15

morphology ☒ p ☐ sf ☐ cf ☐ mf ☐ w *size (μm)*

Processing

primary Stir Casting

specifics

secondary ☐ Casting ☐ Forging ☐ Surface Treatments ☐ Extrusion ☐ Machining ☐ Thermal Treatments

specifics

joining ☐ Brazing ☐ Mechanical ☒ Welding ☐ Other

recyclable ☒ Yes ☐ No

Properties are quoted from manufacturer as general indicators only.

Properties *isotropic* ☐ Yes ☐ No *tensile*

density (g cm⁻³)

CTE ($x10^6 K^{-1}$) 14.9

thermal cond. ($Wm^{-1}K^{-1}$) 120

electrical cond. (%IACS)

yield strength (MPa)

ultimate strength (MPa)

modulus (GPa)

elongation (%)

Property Advantages ☒ Density ☐ Fatigue ☒ Stiffness ☒ Strength ☒ Thermal Expansion ☒ Thermal Conductivity ☒ Wear ☒ Weldability ☐ Other

Applications sailboat masts and other lightweight structural components

Products

Sources Millennium Materials, Inc (2000).

Millennium Materials, Inc (MM)

material (company designation): Carolite®

Material A356 / SiB6 / 40p

matrix Al alloy

reinforcement SiB6 *volume percent* 40

morphology ☒ p ☐ sf ☐ cf ☐ mf ☐ w *size (μm)*

Processing

primary Stir Casting

specifics

secondary ☐ Casting ☐ Forging ☐ Surface Treatments ☐ Extrusion ☐ Machining ☐ Thermal Treatments

specifics

joining ☐ Brazing ☐ Mechanical ☒ Welding ☐ Other

recyclable ☒ Yes ☐ No

Properties are quoted from manufacturer as general indicators only.

Properties *isotropic* ☐ Yes ☐ No *tensile*

density (g cm^{-3}) *yield strength (MPa)*

CTE (x10^{6} K^{-1}) 9.7 *ultimate strength (MPa)*

thermal cond. (Wm^{-1}K^{-1}) 95 *modulus (GPa)* 155

electrical cond. (%IACS) *elongation (%)*

Property Advantages ☒ Density ☒ Strength ☒ Wear ☐ Fatigue ☒ Thermal Expansion ☒ Weldability ☒ Stiffness ☒ Thermal Conductivity ☐ Other

Applications advanced media substrates (disks for hard drives)

Products

Sources Millennium Materials, Inc (2000).

Motorola, Inc

Semiconductor Products Sector
East McDowell Road, Mail Stop A-127
Phoenix AZ 85008 USA

tel ++ 1 (602) 244 6137
fax ++
email mediacenter@motorola.com
web address www.motorola.com

Motorola is a global company involved in a number of sectors including telecommunications and microelectronics. Motorola has made use of MMC for electronic substrates in satellites in collaboration with Lanxide Electronic Components (LEC), now operated by dmc2 Electronic Components (DEC). The Power Products Division of Motorola also uses MMC for heat spreaders in power modules such as IGBT (Insulated Gate Bipolar Transistors).

MSE Technology Applications, Inc (MSE, Inc)

PO Box 4078

Butte MT 59702-4078 USA

tel ++ 1 (406) 494 7100
fax ++ 1 (406) 494 7230
email contact@mse-ta.com
web address www.mse-ta.com

MSE Technology Applications (also known as MSE, Inc) controls the Lanxide technologies through its High Performance Materials Group in Newark DE. The Cast Composites Metal Matrix (CCMM) materials group in Butte MT has developed centrifugal casting processes for producing functionally graded MMC. High performance brake applications are being developed and marketed in collaboration with Swan Metal Composites and the tooling provider sp3.

Olin Aegis
50 Welby Road
New Bedford MA 02745 USA
tel ++ 1 (508) 998 3141
fax ++ 1 (508) 995 7315
email
web address www.olinaegis.com

Olin Aegis markets a range of advanced microelectronic packaging solutions and Al-SiC materials are integrated into a portion of this product line. Al-SiC packages can be plated and assembled into components by Olin Aegis to meet demanding specifications. MMC from the Al-SiC family come primarily from CPS, or for specific applications from PCC-AFT and Ametek. These materials are typically pressure-infiltrated and contain 63 to 68 vol. % silicon carbide.

Hermetic Fibre-Optic packaging solutions are available in a number of configurations, and include opto-electronics components, receiver assemblies and laser diode transmitters. Advanced electronic packages, marketed under the name AEPAC, are available from a number of materials including the Al-SiC family, and can be hermetically encapsulated with plug-ins, can be welded and are compatible with existing sealing systems. Applications have been concentrated in the military and space sectors, however civilian products are emerging.

Omni-Lite Industries
17210 Edwards Road

Cerritos CA 90703 USA

tel ++ 1 (562) 404 8510
fax ++ 1 (562) 926 6913
email d.grant@omni-lite.com
web address www.omni-lite.com

A variety of precision automotive, aerospace and commercial components are manufactured by Omni-Lite utilising computer-controlled cold forging. This cold forging technology is compatible with MMC materials and allows the production of up to 300 components per minute with high tolerance and little material waste. Omni-Lite carefully guards its technology because it has not yet been patented, thus few details are available with regard to which components are produced from MMC. Nevertheless, MMC spikes for track shoes from Omni-Lite have been widely publicised and Omni-Lite dominates the market for track spikes at the professional level.

Orbcomm

21700 Atlantic Boulevard

Dulles VA 20166 USA

tel ++	1 (703) 433 6300 // 1 (800) 672 2666
fax ++	1 (703) 433 6868
email	sales@orbcomm.com
web address	www.orbcomm.com

Orbcomm operates a series of low-orbit satellites that allow customers to track, monitor and manage remote assets from anywhere in the world. These satellites benefit from the use of AlBeMet® 162 material from Brush Wellman for several structural elements. This material exhibits the required specific stiffness needed to survive the launch of the Pegasus spacecraft. The primary structural ring consists of three arched panels of AlBeMet® sheets bonded to a 5052 aluminium alloy honeycomb. These are fastened by the separate brackets, also of AlBeMet®, to provide stiffness and strength for holding separation bolts and shear bearings. The high specific strength is necessary to resist these transient forces upon take-off. The ductility is deemed important for accommodating shock loading at the interface between the non-explosive bolts and the spacecraft, and release of preload energies on the bolts. Two additional applications include four gusset panels to provide rigidity to the structure, and an adhesion bonded AlBeMet® tube to support the antenna boom segments.

PCC Advanced Forming Technology (PCC-AFT)

No. 20
7040 Weld County Road
Longmont CO 80504 USA

tel ++ 1 (303) 833 6000
fax ++ 1 (303) 833 6464
email shanb@pcc-aft.com
web address www.pcc-aft.com

Precision Castparts Corporation (PCC), a large precision casting company, purchased a pressure-infiltration company in Pittsburgh, at the time called P-Cast, and renamed the unit Advanced Forming Technology. PCC-AFT manufactures complex metal components and products, including MMC from the Al-SiC family for electronic packaging.

PCC-AFT primarily serves the defence industry, but it has seen expansion into commercial sectors, particularly for IGBT (Insulated Gate Bipolar Transistor) baseplates. A diverse range of products have been demonstrated by PCC-AFT utilising its patented Pressure Infiltration Casting (PIC™) technology; examples include heat spreaders and carriers, in addition to complex net shaped components such as heatsinks with pin/fin surfaces. PCC-AFT's infiltrated composite components contain 30 to 74 vol. % silicon carbide depending on the requirements of the application and exhibit a coating of unreinforced aluminium over the entire surface of the component. This layer of metal without reinforcement is a patented feature of PCC-AFT's technology that allows parts to be plated and finished as though from bulk metal.

PCC-AFT reports record levels of MMC sales and operating efficiency, providing an optimistic outlook for the future of its AlSiC™ materials in electronic packaging applications. PCC-AFT also claims to have the capability of manufacturing silicon carbide-reinforced copper components for electronic packaging.

PCC Advanced Forming Technology (PCC-AFT)
material (company designation): AlSiC™

Material Al / SiC / 63p

matrix unspecified grade Al
reinforcement SiC *volume percent* 63
morphology ☒ p ☐ sf ☐ cf ☐ mf ☐ w *size (μm)*

Processing
primary Infiltration
specifics PIC™ (patented Pressure Inifltration Casting process)

secondary ☐ Casting ☐ Forging ☒ Surface Treatments ☐ Extrusion ☒ Machining ☐ Thermal Treatments
specifics plating, anodizing, chemical film

joining ☒ Brazing ☐ Mechanical ☐ Welding ☒ Other
recyclable ☐ Yes ☐ No

Properties are quoted from manufacturer as general indicators only.

Properties *isotropic* ☐ Yes ☐ No *tensile*
density (g cm⁻³) 3.01 *yield strength (MPa)*
CTE (x10⁶ K⁻¹) 8.3 (20-100°C) *ultimate strength (MPa)* 253
thermal cond. (Wm⁻¹K⁻¹) *modulus (GPa)* 220
electrical cond. (%IACS) *elongation (%)*

Property Advantages ☒ Density ☒ Strength ☐ Wear ☐ Fatigue ☒ Thermal Expansion ☐ Weldability ☐ Stiffness ☒ Thermal Conductivity ☐ Other

Applications thermal cores, chip carriers, heatsinks, electronic and microwave packages

Products

Sources PCC Advanced Forming Technology (1998).

PCC Advanced Forming Technology (PCC-AFT)
material (company designation): AlSiC™

Material Al / SiC / 68p

matrix unspecified grade Al

reinforcement SiC *volume percent* 68

morphology ☒ p ☐ sf ☐ cf ☐ mf ☐ w *size (μm)*

Processing

primary Infiltration

specifics PIC™ (patented Pressure Iniflltration Casting process)

secondary ☐ Casting ☐ Forging ☒ Surface Treatments ☐ Extrusion ☒ Machining ☐ Thermal Treatments

specifics plating, anodizing, chemical film

joining ☒ Brazing ☐ Mechanical ☐ Welding ☒ Other

recyclable ☐ Yes ☐ No

Properties are quoted from manufacturer as general indicators only.

Properties *isotropic* ☐ Yes ☐ No *tensile*

density (g cm^{-3}) 3.03 *yield strength (MPa)*

CTE (x10^6 K^{-1}) 7.4 (20-100°C) *ultimate strength (MPa)* 207

thermal cond. (Wm^{-1}K^{-1}) *modulus (GPa)* 255

electrical cond. (%IACS) *elongation (%)*

Property Advantages ☒ Density ☐ Fatigue ☐ Stiffness ☒ Strength ☒ Thermal Expansion ☒ Thermal Conductivity ☐ Wear ☐ Weldability ☐ Other

---

Applications thermal cores, chip carriers, heatsinks, electronic and microwave packages

Products

Sources PCC Advanced Forming Technology (1998).

PCC Advanced Forming Technology (PCC-AFT)
material (company designation): AlSiC™

Material Al / SiC / 69p

matrix unspecified grade Al
reinforcement SiC *volume percent* 69
morphology ☒ p ☐ sf ☐ cf ☐ mf ☐ w *size (μm)*

Processing
primary Infiltration
specifics PIC™ (patented Pressure Inifltration Casting process)

secondary ☐ Casting ☐ Forging ☐ Surface Treatments
☐ Extrusion ☐ Machining ☐ Thermal Treatments
specifics plating, anodizing, chemical film

joining ☒ Brazing ☐ Mechanical ☐ Welding ☒ Other
recyclable ☐ Yes ☐ No

Properties are quoted from manufacturer as general indicators only.

Properties *isotropic* ☐ Yes ☐ No *tensile*
density (g cm⁻³) 3.02 *yield strength (MPa)*
CTE ($x10^6 K^{-1}$) 6.9 *ultimate strength (MPa)*
thermal cond. ($Wm^{-1}K^{-1}$) *modulus (GPa)* 225
electrical cond. (%IACS) *elongation (%)*

Property Advantages ☒ Density ☒ Strength ☐ Wear
☐ Fatigue ☒ Thermal Expansion ☐ Weldability
☐ Stiffness ☒ Thermal Conductivity ☐ Other

Applications thermal cores, chip carriers, heatsinks, electronic and microwave packages

Products

Sources Sonuparlak, B.; Andrews, D.J. (1998).

Poeton Industries Ltd
Eastern Avenue
Gloucester GL4 3DN UK

tel ++ 44 (0)1452 300500
fax ++ 44 (0)1452 500400
email sales@poeton.co.uk
web address http://www.poeton.co.uk/

With over one hundred years of experience, the Poeton Group of Companies are active in Surface Engineering with coatings and surface treatments to reduce wear, friction and corrosion. Specialising in electroplating, anodising, plasma electrolytic oxidation, electroless nickel, thermal spraying and polymer coatings, the company offers a world wide service to all the major industrial sectors.

The company offers a wide range of coatings, several of which are metal matrix composites. Examples are:

- Apticote 450, which is a self-lubricating composite of PTFE embedded in a matrix of electroless nickel;

-the Apticote 800 range of plasma flame sprayed coatings, which includes an aluminium alloy graphite composite (Apticote 800/67), and a nickel-chrome ceramic composite (Apticote 850);

- Apticote 2000, a nickel-based coating reinforced with typically from 12 to 16% hard ceramic particles for engine block applications.

(information updated in Feb 2003)

Polese Company
10121 Carroll Canyon Road
San Diego CA 92131 USA

tel ++ 1 (858) 271 1993
fax ++ 1 (858) 271 4376
email sales@polese.com
web address www.polese.com

Polese, the Microelectronic Packaging Group of SEMX Corporation, manufactures and markets a range of products for thermal management applications in the communications and computer industries. They offer a broad range of materials for thermal management applications including materials from the Al-SiC family, as well as W/Mo-Cu materials and its patented Silvar™ materials.

The Al-SiC materials became established at Polese through internal development of low-cost sheet and powder metallurgical processes, in addition to patented vacuum casting technology (and pilot plant) acquired from Alcoa. Polese offers three basic grades of Al-SiC materials primarily for thermal management applications. Series 700 has the lowest thermal expansion and the highest thermal conductivity, presumably because it has the largest volume fraction of silicon carbide. Series 900 is marketed as an excellent cost/performance tradeoff, while series 1200 is promoted as the most economical composition. Polese markets these materials as "low cost, lightweight thermal management options" for applications in lids for Flip Chip power devices, microwave carriers and heat sinks, power module baseplates, and microelectronic package bases. The intrinsic properties of these materials, namely high stiffness, low density, low thermal expansion, and good thermal conductivity, make them suitable for wave-guide applications, such as microwave carriers, over a range of temperatures.

Copper-tungsten, copper-molybdenum, and silver-iron materials are also offered by Polese for applications similar to those exploited by the Al-SiC family of materials. A copper-tungsten material from Polese is used in the Ti Bubble 2™ irons from Taylor Made Golf; this is an expansion into a new sector for Polese, offering potential for future growth.

Polese Company

material (company designation): AlSiC series 700

Material Al / SiC / p

matrix unspecified grade Al

reinforcement SiC *volume percent*

morphology ☒ p ☐ sf ☐ cf ☐ mf ☐ w *size (μm)*

Processing

primary Infiltration

specifics net-shape vacuum casting

secondary ☐ Casting ☐ Forging ☐ Surface Treatments ☐ Extrusion ☐ Machining ☐ Thermal Treatments

specifics

joining ☐ Brazing ☐ Mechanical ☐ Welding ☐ Other

recyclable ☐ Yes ☐ No

Properties are quoted from manufacturer as general indicators only.

Properties *isotropic* ☐ Yes ☐ No *tensile*

density (g cm^{-3}) 3.02 *yield strength (MPa)*

CTE (x10^{6} K^{-1}) 6-8 *ultimate strength (MPa)*

thermal cond. (Wm^{-1}K^{-1}) 210 *modulus (GPa)* 224

electrical cond. (%IACS) *elongation (%)*

Property Advantages ☒ Density ☐ Strength ☐ Wear ☐ Fatigue ☒ Thermal Expansion ☐ Weldability ☐ Stiffness ☒ Thermal Conductivity ☐ Other

Applications thermal management for electronics: flip chip lids, microwave carriers, power module bases, microelectronic package bases

Products

Sources Polese (1999).

Polese Company

material (company designation): AlSiC series 900

Material Al / SiC / p

matrix unspecified grade Al

reinforcement SiC *volume percent*

morphology ☒ p ☐ sf ☐ cf ☐ mf ☐ w *size (μm)*

Processing

primary

specifics net-shape vacuum casting

secondary ☐ Casting ☐ Forging ☐ Surface Treatments ☐ Extrusion ☐ Machining ☐ Thermal Treatments

specifics

joining ☐ Brazing ☐ Mechanical ☐ Welding ☐ Other

recyclable ☐ Yes ☐ No

Properties are quoted from manufacturer as general indicators only.

Properties *isotropic* ☐ Yes ☐ No *tensile*

density (g cm^{-3}) 2.93 *yield strength (MPa)*

CTE (x10^{6} K^{-1}) 8-10 *ultimate strength (MPa)*

thermal cond. (Wm^{-1}K^{-1}) 160 *modulus (GPa)* 118

electrical cond. (%IACS) *elongation (%)*

Property Advantages ☒ Density ☐ Strength ☐ Wear ☐ Fatigue ☒ Thermal Expansion ☐ Weldability ☐ Stiffness ☒ Thermal Conductivity ☐ Other

Applications thermal management for electronics: flip chip lids, microwave carriers, power module bases, microelectronic package bases

Products

Sources Polese (1999).

Polese Company

material (company designation): AlSiC series 1200

Material Al / SiC / p

matrix unspecified grade Al

reinforcement SiC *volume percent*

morphology ☒ p ☐ sf ☐ cf ☐ mf ☐ w *size (μm)*

Processing

primary

specifics net-shape vacuum casting

secondary ☐ Casting ☐ Forging ☐ Surface Treatments ☐ Extrusion ☐ Machining ☐ Thermal Treatments

specifics

joining ☐ Brazing ☐ Mechanical ☐ Welding ☐ Other

recyclable ☐ Yes ☐ No

Properties are quoted from manufacturer as general indicators only.

Properties *isotropic* ☐ Yes ☐ No *tensile*

density ($g\ cm^{-3}$) 2.79 *yield strength (MPa)*

CTE ($x10^6\ K^{-1}$) 12-14 *ultimate strength (MPa)*

thermal cond. ($Wm^{-1}K^{-1}$) 170 *modulus (GPa)* 110

electrical cond. (%IACS) *elongation (%)*

Property Advantages ☒ Density ☐ Fatigue ☐ Stiffness ☐ Strength ☒ Thermal Expansion ☒ Thermal Conductivity ☐ Wear ☐ Weldability ☐ Other

Applications thermal management for electronics: flip chip lids, microwave carriers, power module bases, microelectronic package bases

Products

Sources Polese (1999).

Porsche

Germany

tel ++
fax ++
email
web address www.porsche.com

Porsche is one of the largest manufacturers of high-performance sports cars. With the introduction of the Boxster in 1997, Porsche embraced MMC technology. The 2.5 litre six-cylinder engine has cylinder bores selectively reinforced with silicon particles, a manufacturing technology developed by Kolbenschmidt Pierburg (KS Aluminium Technologie). The success of the technology led to the implementation of the composite, called LOKASIL®, in the 3.4 litre flat six for the new Carrera 911.

Powdermet Inc

Unit A
9960 Glenoaks Boulevard
Sun Valley CA 91352 USA

tel ++ 1 (818) 768 6420
fax ++ 1 (818) 768 6012
email powdermet@earthlink.net
web address www.powdermetinc.com

Powdermet produces a diverse range of coated reinforcement particulate by a patented Recirculating Fast Fluidized Bed Chemical Vapor Deposition processing technology (RFFBCVD). Bulk MMC can be produced from these metal-coated reinforcements by powder metallurgical methods or thermal spraying. Some of the available matrix metals include aluminium, nickel, copper, silver, and silicon; reinforcements include SiC, TiB_2, B_4C, diamond, graphite and pitch fibres. Powdermet has been awarded a Phase I SBIR from the Ballistic Missile Defense Organization, with an aim to improve MMC materials and forming methods by powder forging and thermal spraying. Powdermet will focus on processing of aluminium-coated silicon carbide particulate called LiteAloy™.

The Small Business Innovation Research (SBIR) and the Small Business Technology Transfer (STTR) programmes are funded by the US government to assist small businesses in expanding research and development.

Pratt and Whitney
400 Main Street
East Hartford CT 06108 USA

tel ++ 1 (860) 565 4321
fax ++ 1 (860) 565 3987
email
web address www.pratt-whitney.com

Pratt and Whitney (PW) manufactures a range of power and propulsion systems, including large turbofan jet engines and parts for both commercial and military aircraft, small gas turbines, space and missile propulsion systems, and land-based power systems. PW was involved in the Title III program (sponsored by the US government) that led to the development of discontinuously reinforced aluminium (sometimes called DRA) for use in the PW4000 series engines; these engines power a large proportion of Boeing 777 aircraft. Fan Exit Guide Vanes (FEGV) on the bypass section of the fan are produced from 6092/SiC/17.5 supplied by DWA Aluminum Composites. This MMC component replaces a polymer matrix composite that suffered erosion problems and poor resistance to impact, thus requiring frequent replacement. A financial incentive for this substitution also existed, since manufacturing of the component from polymer matrix composite is costly, both in terms of labour and materials. The use of MMC is projected to save PW 100 million USD in acquisition costs over 10 years.

Precorp, Inc
1962 North Chapel Drive

Spanish Fork UT 84660 USA

tel ++ 1 (801) 798 5425
fax ++ 1 (801) 798 4698
email johnc@Precorp.net
web address www.precorp.net

Precorp offers a full line of polycrystalline diamond (PCD) tooling, called DiaEdge®, for machining MMC and other materials. Precorp initially provided tooling for a research programme with Lockheed to machine aluminium reinforced with 45 vol. % silicon carbide particles (Al/SiC/45p). DiaEdge® tooling was extensively studied for machining Duralcan materials from Alcan and studies show that the cost of drilling with PCD can be as low as a quarter of that using carbide. Precorp tooling was recommended by Alcan.

ProEngCo Tooling AB

Fabriksvägen 2

Staffanstorp 24534 Sweden

tel ++ 46 (46) 189 008
fax ++ 46 (46) 189 009
email anders.nilsson@proengco.se
web address www.proengco.se (go to Proengco Tooling AB)

ProEngCo has developed and obtained a patent for a method of casting, joining and pressing of materials utilising high-frequency vibrations, a process known as Vibrocast™. High-frequency vibration and acoustic waves stimulate the melt leading to improved microstructures (grain refinement and homogeneity) and hence improved mechanical properties. In composite casting, this process improves infiltration and the wetting characteristics, allowing infiltration at relatively low pressures, below 2 MPa for aluminium reinforced by 40 vol. % Saffil®. Near-net forming of diverse microstructures (aluminium and magnesium alloys with Al_2O_3, SiC, C and metal fibres) are possible with this process.

The developmental work was partly financed by the Swedish Defence Materials Administration (FMV), although the technology is available for commercial exploitation.

QED Extrusion Developments

11808 Rancho Bernardo Road #123

San Diego CA 92128 USA

tel ++	1 (760) 432 9532
fax ++	1 (760) 432 9740
email	billdixon@qedextrusion.com
web address	www.bricad.com/clients.html

QED is a company specializing in extrusion, offering product and process development services to both the extrusion house and the end-user. QED can manufacture prototypes and small batches of aluminium-based MMC. QED's experience includes extrusion of MMC materials in both the UK, Germany and Italy, manufacture of initial orders for MMC for use in a wide variety of markets, including automotive, aerospace, recreational, electronics. Commercialization has led to large volume production of MMC extrusions and rolled product in excess of 300,000 lbs per year.

(Updated Feb. 2003)

Saffil Ltd

Pilkington Sullivan Site
Tanhouse Lane
Widnes, Cheshire WA8 0RY UK

tel ++ 44 (0) 151 422 6700
fax ++ 44 (0) 151 422 6701
email info@saffil.com
web address www.saffil.com

Saffil Ltd supplies alumina short-fibre forms (or preforms) to the automotive industry for producing MMCs, particularly for selective reinforcement of engine blocks, pistons and brake parts.

Saffil is owned by the Dyson Group plc, 381 Fulwood Road, Sheffield S10 3GB, UK (www.dyson-group.com).

(Updated Feb 2003)

Sageborn Bicycles

Via S. Patrizio 23

Bassano del Grappa VI 36061 Italy

tel ++ 39 (424) 56 64 45
fax ++ 39 (424) 56 68 04
email sage@nsoft.it
web address www.nsoft.it/sageborn

Sageborn Bicycles market the Ceramical 100 bicycle frames produced from ceramic particle-reinforced 7020 aluminium alloy. Frames for both road racing and off-road use are available. Due to the higher specific strength and stiffness of the Ceramical 100 tubing compared to titanium and aluminium alloys, the tube diameter is not substantially oversized, which has the virtue of avoiding an overly rigid ride.

SatCon Technology Corporation
161 First Street

Cambridge MA 02142 USA

tel ++ 1 (617) 661 0540
fax ++ 1 (617) 661 3373
email flodin@satc.com
web address www.satcon.com

SatCon manufactures power and energy management products. In developing highly efficient induction motors that operate at 100,000 rpm, SatCon has found that conventional copper end rings would fail. To counter this problem, SatCon uses a continuous fibre-reinforced metal (CFRM) retainer ring developed in collaboration with 3M. The composite retainer ring contains 68 vol. % hoop-wound continuous fibres embedded in an aluminium matrix. The retainer rings are press-fit to the copper shaft. The rotor has been successfully tested to 100,000 rpm, the speed at which the turbine becomes unstable. The estimated speed for yield of the metal matrix composite is 125,000 rpm.

The continuous fibre aluminium matrix composite material was chosen over other high performance materials for several reasons. While the material may prove more costly than graphite-epoxy flywheels, the polymer matrix composite would require a larger support structure. 3M's continuous fibre-reinforced aluminium composite is thus better for space-limited rotor applications, and CFRM will not degas, facilitating its operation in vacuum. Additionally, being non ferrous, CFRM can operate in magnetic fields.

Siemens AG

Siemens AG, CT MS 4

Germany

tel ++ 49 89 636 45163
fax ++ 49 89 636 46376
email guy.lefranc@siemens.com
web address www.siemens.com

Siemens is a world leader in the electronics and electrical engineering sectors. The trend in this arena has been towards miniaturisation resulting in increased power densities and problems of thermal management in electronic packaging. The mismatch in thermal expansion between the packaging and the device results in thermal stresses in electronic packages. Rupture due to thermal fatigue in the solder joint that typically bonds these layers together causes system failure. Thus, matching the coefficient of thermal expansion (CTE) of these layers is an important design consideration, an this requires baseplates with a low CTE to match that of the ceramic substrate. The baseplate also acts as a heat-spreader removing heat from the electronic device, thus a high thermal conductivity is desirable. Siemens and its former subsidiary EUPEC* have developed designs for IGBT (Insulated Gate Bipolar Transistor) modules with baseplates from the Al-SiC family of materials; several thousand modules have been manufactured. With high volume fractions of silicon carbide, Al-SiC materials can have a thermal conductivity of more than 200 $Wm^{-1}K^{-1}$, a value greater than the matrix alloy. Development of silicon carbide-reinforced copper materials has also been reported.

*(eupec, european power-semiconductor and electronics company, is now an Infineon Technologies subsidiary).

(updated Feb. 2003)

sp3

505 East Evelyn Avenue

Mountain View CA USA

tel ++ 1 (650) 966 0630
fax ++
email sales@sp3inc.com
web address www.sp3inc.com

sp3 is one of the most experienced tool makers active in the MMC industry. sp3 offers a complete line of DIAbide™ tooling: endmills, moldmills, inserts, drills, profiling inserts, routers, grooving, cutoff and threading inserts, and saws. DIAbide™ is tooling carbide coated with an adherent diamond film by chemical vapour deposition (CVD). Extensive testing on Duralcan material (from Alcan) has shown that the lifetime of DIAbide™ tooling is up to seven times greater than polycrystalline diamond (PCD) in milling operations and up to three time greater in turning operations.

DIAbide™ tooling has been used by MSE, Inc. and in collaboration with Swan Metal Composites to fabricate brake discs and other MMC components with substantial increase in performance over uncoated carbide tooling. Longer tool life is reported to increase productivity and lower machining costs, as well as to facilitate the manufacture of MMC components that could not be produced otherwise.

Sumitomo Electric USA, Inc

Suite 200
21221 S. Western Avenue
Torrance CA 90501 USA

tel ++ 1 (310) 782 0227
fax ++ 1 (310) 782 0211
email
web address

Sumitomo Electric manufactures silicon carbide particle-reinforced aluminium materials by direct forming of powder material into a near-net shape component. The aluminium-matrix alloy powder and the silicon carbide particulate are mixed with an additive and then compacted and sintered to full density and net shape. The particulate concentration can be reliably varied over a large range with this processing method; Sumitomo Electric offers MMC with anywhere between 30 and 70 vol. % SiC. These materials are aimed at the electronic packaging markets, for example as heatsinks for the semiconductor industry.

Sumitomo Electric USA, Inc

material (company designation):

Material Al / SiC / 30p

matrix unspecified grade Al

reinforcement SiC *volume percent* 30

morphology ☒ p ☐ sf ☐ cf ☐ mf ☐ w *size (μm)*

Processing

primary Powder Metallurgy

specifics

secondary ☐ Casting ☐ Forging ☐ Surface Treatments ☐ Extrusion ☐ Machining ☐ Thermal Treatments

specifics

joining ☐ Brazing ☐ Mechanical ☐ Welding ☐ Other

recyclable ☐ Yes ☐ No

Properties are quoted from manufacturer as general indicators only.

Properties *isotropic* ☐ Yes ☐ No *tensile*

density (g cm^{-3}) 2.78 *yield strength (MPa)*

CTE (x10^{6} K^{-1}) 18 *ultimate strength (MPa)* 98

thermal cond. (Wm^{-1}K^{-1}) 200 *modulus (GPa)*

electrical cond. (%IACS) *elongation (%)* 6.8

Property Advantages ☐ Density ☐ Strength ☐ Wear ☐ Fatigue ☒ Thermal Expansion ☐ Weldability ☐ Stiffness ☒ Thermal Conductivity ☐ Other

Applications thermal management for electronics

Products

Sources

Sumitomo Electric USA, Inc

material (company designation):

Material Al / SiC / 50p

matrix unspecified grade Al

reinforcement SiC *volume percent* 50

morphology ☒ p ☐ sf ☐ cf ☐ mf ☐ w *size (μm)*

Processing

primary Powder Metallurgy

specifics

secondary ☐ Casting ☐ Forging ☐ Surface Treatments ☐ Extrusion ☐ Machining ☐ Thermal Treatments

specifics

joining ☐ Brazing ☐ Mechanical ☐ Welding ☐ Other

recyclable ☐ Yes ☐ No

Properties are quoted from manufacturer as general indicators only.

Properties *isotropic* ☐ Yes ☐ No *tensile*

density ($g\ cm^{-3}$) 2.70 *yield strength (MPa)*

CTE ($x10^6\ K^{-1}$) 12 *ultimate strength (MPa)* 147

thermal cond. ($Wm^{-1}K^{-1}$) 180 *modulus (GPa)*

electrical cond. (%IACS) *elongation (%)* 0.5

Property Advantages ☐ Density ☐ Strength ☐ Wear ☐ Fatigue ☒ Thermal Expansion ☐ Weldability ☐ Stiffness ☒ Thermal Conductivity ☐ Other

Applications thermal management for electronics

Products

Sources

Sumitomo Electric USA, Inc

material (company designation):

Material Al / SiC / 70p

matrix unspecified grade Al

reinforcement SiC *volume percent* 70

morphology ☒ p ☐ sf ☐ cf ☐ mf ☐ w *size (μm)*

Processing

primary Powder Metallurgy

specifics

secondary ☐ Casting ☐ Forging ☐ Surface Treatments ☐ Extrusion ☐ Machining ☐ Thermal Treatments

specifics

joining ☐ Brazing ☐ Mechanical ☐ Welding ☐ Other

recyclable ☐ Yes ☐ No

Properties are quoted from manufacturer as general indicators only.

Properties *isotropic* ☐ Yes ☐ No *tensile*

density (g cm^{-3}) 2.60 *yield strength (MPa)*

CTE (x10^{6} K^{-1}) 8 *ultimate strength (MPa)* 78

thermal cond. (Wm^{-1}K^{-1}) 150 *modulus (GPa)*

electrical cond. (%IACS) *elongation (%)* 0.2

Property Advantages ☐ Density ☐ Strength ☐ Wear ☐ Fatigue ☒ Thermal Expansion ☐ Weldability ☐ Stiffness ☒ Thermal Conductivity ☐ Other

Applications thermal management for electronics

Products

Sources

Swan Metal Composites
17438 174th Avenue N.E.
Woodinville WA 98072 USA

tel ++ 1 (425) 398 0588
fax ++ 1 (425) 398 0588
email tc_mmc@msn.com
web address

Swan Metal Composites, Inc. produces directly cast functionally graded and selectively reinforced metal composite products. Swan Metal Composites owns broad intellectual property rights for manufacture of functionally graded and selectively reinforced metal composites, in addition to holding a sublicense to US Navy intellectual property. Commercial applications have included brake rotors for high performance and racing vehicles, industrial equipment and electronics. Use of a functionally graded MMC (silicon carbide particle reinforced aluminum) friction ring for the front brakes of motorcycles has allowed a reduction in overall assembly mass of 50% relative to ferrous based assemblies. The weight reduction significantly improves handling through the reduction of unsprung and rotational mass. These benefits have been realized for recent models of sport motorcycles with successful fitment to liter class racing motorcycles manufactured by Ducati and Suzuki, for which MMC brake rotors and associated hardware have been produced for substitution of original parts. Swan Metal Composites has also collaborated in the development of Sp3's DIAbideTM tooling materials, with the aim of increasing machining feed rate and improving tool durability.

Referenced intellectual property includes U.S. Patent No. 5,980,792, "Particulate Field Distributions in Centrifugally Cast Composites", including Australian Patent No. 732,289. Additional European patents pending. Sublicense references U.S. Patent No. 5,025,849, "Centrifugal Casting of Composites", held by the US Navy.

Functionally graded, internally ventilated, metal composite brake rotors have also been successfully developed for 250cc Unlimited class Racing Karts (year 2002 work). Current work includes the manufacture of functionally graded aluminum-bronze / WCp components for improved wear properties.

(updated Feb. 2003)

Talon Composites, Inc
25677 Paseo de la Paz
San Juan Capistrano CA - 92675 U.S.A
tel ++ 1 949 248 0005 / 1 949 412 1077 (24h)
fax ++ 1 949 248 9971
email r.carden@cox.net
web address www.taloncomposites.com

Talon Composites produces advanced materials, including a variety of particle-reinforced aluminium composites with the trade name Talbor®. Powder metallurgy is used to produce materials with 5 to 40 vol. % reinforcement as extruded billet or forged stock. Cast ingot is also produced with as much 15 vol. % reinforcement. Talbor® is tailorable for specific requirements both in terms of the matrix and reinforcement. The aluminium matrix can be wrought heat-treatable 2000, 6000 or 7000 series, nonheat-treatable 5000 series, or A300 series casting alloy. Silicon carbide, boron carbide and alumina are currently employed as reinforcing phases; silicon hexaboride remains an option for future development.

Composites containing SiC are not easily weldable because porosity develops in the heat-affected zone due to degassing from the silicon carbide particulate. The B_4C reinforcement does not degas, thus weldability of Talbor® reinforced with boron carbide is determined by the base alloy. The SiC-reinforced aluminium, however, is stronger than the B_4C-reinforced aluminium. The Al_2O_3-reinforced aluminium is the strongest in the Talbor® family; in addition, since alumina is stable when in contact with an aluminium melt, it can be processed as a liquid from a greater variety of alloys than the SiC compositions.

Talon is pursuing both commercial and nuclear markets, the latter served by the B_4C compositions due to the neutron absorbing properties of boron carbide. These compositions are thus suited for dry cask storage and wet rack storage of nuclear materials in addition to a host of nonstorage applications such as reactor shielding. The market for nuclear waste storage materials appears to be dominated by a few companies, and Talon expects that a significant demand for neutron absorbing materials will be created by projects such as the US federal storage site.

Talon Composites, Inc

material (company designation): Talbor® ΔB5 (Delta B5)

Material 6092 / B4C / 5p / T6

matrix Al alloy

reinforcement B4C *volume percent* 5

morphology ☒ p ☐ sf ☐ cf ☐ mf ☐ w *size (μm)*

Processing

primary Powder Metallurgy

specifics

secondary ☐ Casting ☒ Forging ☐ Surface Treatments ☒ Extrusion ☐ Machining ☐ Thermal Treatments

specifics

joining ☐ Brazing ☐ Mechanical ☐ Welding ☐ Other

recyclable ☐ Yes ☐ No

Properties are quoted from manufacturer as general indicators only.

Properties *isotropic* ☐ Yes ☐ No *tensile*

density (g cm^{-3}) 2.681 *yield strength (MPa)* 345

CTE (x10^{6} K^{-1}) 22.5 *ultimate strength (MPa)* 414

thermal cond. (Wm^{-1}K^{-1}) 173 *modulus (GPa)* 77.9

electrical cond. (%IACS) *elongation (%)* 8

Property Advantages ☒ Density ☒ Strength ☒ Wear ☒ Fatigue ☐ Thermal Expansion ☒ Weldability ☒ Stiffness ☐ Thermal Conductivity ☐ Other

Applications structural neutron absorption applications; sport equipment tubing; gears and pistons in engine applications; guide vanes for military aircraft; wrist pins and skid plates

Products

Sources

Talon Composites, Inc

material (company designation): Talbor® ΔB10 (Delta B10)

Material 6092 / B4C / 10p / T6

matrix Al alloy

reinforcement B4C *volume percent* 10

morphology ☒ p ☐ sf ☐ cf ☐ mf ☐ w *size (μm)*

Processing

primary Powder Metallurgy

specifics

secondary ☐ Casting ☒ Forging ☐ Surface Treatments ☒ Extrusion ☐ Machining ☐ Thermal Treatments

specifics

joining ☐ Brazing ☐ Mechanical ☐ Welding ☐ Other

recyclable ☐ Yes ☐ No

Properties are quoted from manufacturer as general indicators only.

Properties *isotropic* ☐ Yes ☐ No *tensile*

density ($g\ cm^{-3}$) 2.675 *yield strength (MPa)* 345

CTE ($x10^{6}\ K^{-1}$) 21.1 *ultimate strength (MPa)* 414

thermal cond. ($Wm^{-1}K^{-1}$) 166 *modulus (GPa)* 84.1

electrical cond. (%IACS) *elongation (%)* 5

Property Advantages ☒ Density ☒ Strength ☒ Wear ☒ Fatigue ☐ Thermal Expansion ☒ Weldability ☒ Stiffness ☐ Thermal Conductivity ☐ Other

Applications structural neutron absorption applications; sport equipment tubing; gears and pistons in engine applications; guide vanes for military aircraft; wrist pins and skid plates

Products

Sources

Talon Composites, Inc

material (company designation): Talbor® ΔB15 (Delta B15)

Material 6092 / B4C / 15p / T6

matrix Al alloy

reinforcement B4C *volume percent* 15

morphology ☒ p ☐ sf ☐ cf ☐ mf ☐ w *size (μm)*

Processing

primary Powder Metallurgy

specifics

secondary ☐ Casting ☒ Forging ☐ Surface Treatments ☒ Extrusion ☐ Machining ☐ Thermal Treatments

specifics

joining ☐ Brazing ☐ Mechanical ☐ Welding ☐ Other

recyclable ☐ Yes ☐ No

Properties are quoted from manufacturer as general indicators only.

Properties *isotropic* ☐ Yes ☐ No *tensile*

density (g cm^{-3}) 2.667 *yield strength (MPa)* 345

CTE (x10^{6} K^{-1}) 19.1 *ultimate strength (MPa)* 414

thermal cond. (Wm^{-1}K^{-1}) 156 *modulus (GPa)* 92.4

electrical cond. (%IACS) *elongation (%)* 4

Property Advantages ☒ Density ☒ Strength ☒ Wear ☒ Fatigue ☐ Thermal Expansion ☒ Weldability ☒ Stiffness ☐ Thermal Conductivity ☐ Other

Applications structural neutron absorption applications; sport equipment tubing; gears and pistons in engine applications; guide vanes for military aircraft; wrist pins and skid plates

Products

Sources

Talon Composites, Inc

material (company designation): Talbor® ΔB20 (Delta B20)

Material	6092 / B4C / 20p / T6
matrix	Al alloy
reinforcement	B4C — *volume percent* 20
morphology	☒ p ☐ sf ☐ cf ☐ mf ☐ w — *size (μm)*

Processing	
primary	Powder Metallurgy
specifics	
secondary	☐ Casting ☒ Forging ☐ Surface Treatments ☒ Extrusion ☐ Machining ☐ Thermal Treatments
specifics	
joining	☐ Brazing ☐ Mechanical ☐ Welding ☐ Other
recyclable	☐ Yes ☐ No

Properties are quoted from manufacturer as general indicators only.

Properties		*isotropic* ☐ Yes ☐ No	*tensile*
density (g cm^{-3})	2.661	*yield strength (MPa)*	345
CTE (x10^{6} K^{-1})	17.5	*ultimate strength (MPa)*	414
thermal cond. (Wm^{-1}K^{-1})	151	*modulus (GPa)*	101.4
electrical cond. (%IACS)		*elongation (%)*	3

Property Advantages	☒ Density	☒ Strength	☒ Wear
	☒ Fatigue	☐ Thermal Expansion	☒ Weldability
	☒ Stiffness	☐ Thermal Conductivity	☐ Other

Applications	structural neutron absorption applications; sport equipment tubing; gears and pistons in engine applications; guide vanes for military aircraft; wrist pins and skid plates
Products	
Sources	

Talon Composites, Inc

material (company designation): Talbor® ΔB25 (Delta B25)

Material 6092 / B4C / 25p / T6

matrix Al alloy

reinforcement B4C *volume percent* 25

morphology ☒ p ☐ sf ☐ cf ☐ mf ☐ w *size (μm)*

Processing

primary Powder Metallurgy

specifics

secondary ☐ Casting ☒ Forging ☐ Surface Treatments ☒ Extrusion ☐ Machining ☐ Thermal Treatments

specifics

joining ☐ Brazing ☐ Mechanical ☐ Welding ☐ Other

recyclable ☐ Yes ☐ No

Properties are quoted from manufacturer as general indicators only.

Properties *isotropic* ☐ Yes ☐ No *tensile*

density (g cm^{-3}) 2.653 *yield strength (MPa)* 345

CTE (x10^{6} K^{-1}) 16.2 *ultimate strength (MPa)* 414

thermal cond. (Wm^{-1}K^{-1}) 145 *modulus (GPa)* 111.0

electrical cond. (%IACS) *elongation (%)* 2

Property Advantages ☒ Density ☒ Strength ☒ Wear ☒ Fatigue ☐ Thermal Expansion ☒ Weldability ☒ Stiffness ☐ Thermal Conductivity ☐ Other

Applications structural neutron absorption applications; sport equipment tubing; gears and pistons in engine applications; guide vanes for military aircraft; wrist pins and skid plates

Products

Sources

Talon Composites, Inc

material (company designation): Talbor® ΔB30 (Delta B30)

Material 6092 / B4C / 30p / T6

matrix Al alloy

reinforcement B4C *volume percent* 30

morphology ☒ p ☐ sf ☐ cf ☐ mf ☐ w *size (μm)*

Processing

primary Powder Metallurgy

specifics

secondary ☐ Casting ☒ Forging ☐ Surface Treatments ☒ Extrusion ☐ Machining ☐ Thermal Treatments

specifics

joining ☐ Brazing ☐ Mechanical ☐ Welding ☐ Other

recyclable ☐ Yes ☐ No

Properties are quoted from manufacturer as general indicators only.

Properties *isotropic* ☐ Yes ☐ No

			tensile
density (g cm^{-3})	2.648	*yield strength (MPa)*	345
CTE (x10^{6} K^{-1})	14.9	*ultimate strength (MPa)*	414
thermal cond. (Wm^{-1}K^{-1})	135	*modulus (GPa)*	120.7
electrical cond. (%IACS)		*elongation (%)*	1

Property Advantages ☒ Density ☒ Strength ☒ Wear ☒ Fatigue ☐ Thermal Expansion ☒ Weldability ☒ Stiffness ☐ Thermal Conductivity ☐ Other

Applications structural neutron absorption applications; sport equipment tubing; gears and pistons in engine applications; guide vanes for military aircraft; wrist pins and skid plates

Products

Sources

Talon Composites, Inc

material (company designation): Talbor® ΔB35 (Delta B35)

Material 6092 / B4C / 35p / T6

matrix Al alloy

reinforcement B4C *volume percent* 35

morphology ☒ p ☐ sf ☐ cf ☐ mf ☐ w *size (μm)*

Processing

primary Powder Metallurgy

specifics

secondary ☐ Casting ☒ Forging ☐ Surface Treatments ☒ Extrusion ☐ Machining ☐ Thermal Treatments

specifics

joining ☐ Brazing ☐ Mechanical ☐ Welding ☐ Other

recyclable ☐ Yes ☐ No

Properties are quoted from manufacturer as general indicators only.

Properties *isotropic* ☐ Yes ☐ No *tensile*

density ($g\ cm^{-3}$) 2.639 *yield strength (MPa)* 345

CTE ($x10^6\ K^{-1}$) 14.2 *ultimate strength (MPa)* 414

thermal cond. ($Wm^{-1}K^{-1}$) 125 *modulus (GPa)* 131.7

electrical cond. (%IACS) *elongation (%)* 0.75

Property Advantages ☒ Density ☒ Strength ☒ Wear ☒ Fatigue ☐ Thermal Expansion ☒ Weldability ☒ Stiffness ☐ Thermal Conductivity ☐ Other

Applications structural neutron absorption applications; sport equipment tubing; gears and pistons in engine applications; guide vanes for military aircraft; wrist pins and skid plates

Products

Sources

Talon Composites, Inc

material (company designation): Talbor® ΔS5 (Delta S5)

Material 6092 / SiC / 5p / T6

matrix Al alloy

reinforcement SiC *volume percent* 5

morphology ☒ p ☐ sf ☐ cf ☐ mf ☐ w *size (μm)*

Processing

primary Powder Metallurgy

specifics

secondary ☐ Casting ☒ Forging ☐ Surface Treatments ☒ Extrusion ☐ Machining ☒ Thermal Treatments

specifics

joining ☐ Brazing ☐ Mechanical ☐ Welding ☐ Other

recyclable ☐ Yes ☐ No

Properties are quoted from manufacturer as general indicators only.

Properties *isotropic* ☐ Yes ☐ No *tensile*

density (g cm^{-3}) 2.717 *yield strength (MPa)* 372

CTE (x10^{6} K^{-1}) 22.5 *ultimate strength (MPa)* 428

thermal cond. (Wm^{-1}K^{-1}) 180 *modulus (GPa)* 76.5

electrical cond. (%IACS) *elongation (%)* 8

Property Advantages ☒ Density ☒ Strength ☒ Wear ☒ Fatigue ☐ Thermal Expansion ☐ Weldability ☒ Stiffness ☐ Thermal Conductivity ☐ Other

Applications structural components with high specific strength and stiffness such as robotics and non-welded tubing assemblies; shafts for sporting equipment; pistons and connecting rods for motorsport; friction components

Products

Sources

Talon Composites, Inc

material (company designation): Talbor® ΔS10 (Delta S10)

Material 6092 / SiC / 10p / T6

matrix Al alloy

reinforcement SiC *volume percent* 10

morphology ☒ p ☐ sf ☐ cf ☐ mf ☐ w *size (μm)*

Processing

primary Powder Metallurgy

specifics

secondary ☐ Casting ☒ Forging ☐ Surface Treatments ☒ Extrusion ☐ Machining ☒ Thermal Treatments

specifics

joining ☐ Brazing ☐ Mechanical ☐ Welding ☐ Other

recyclable ☐ Yes ☐ No

Properties are quoted from manufacturer as general indicators only.

Properties *isotropic* ☐ Yes ☐ No *tensile*

density ($g\ cm^{-3}$) 2.741 *yield strength (MPa)* 386

CTE ($x10^6\ K^{-1}$) 21.1 *ultimate strength (MPa)* 441

thermal cond. ($Wm^{-1}K^{-1}$) 183 *modulus (GPa)* 84.1

electrical cond. (%IACS) *elongation (%)* 6

Property Advantages ☒ Density ☒ Fatigue ☒ Stiffness ☒ Strength ☐ Thermal Expansion ☐ Thermal Conductivity ☒ Wear ☐ Weldability ☐ Other

Applications structural components with high specific strength and stiffness such as robotics and non-welded tubing assemblies; shafts for sporting equipment; pistons and connecting rods for motorsport; friction components

Products

Sources

Talon Composites, Inc

material (company designation): Talbor® ΔS15 (Delta S15)

Material 6092 / SiC / 15p / T6

matrix Al alloy

reinforcement SiC *volume percent* 15

morphology ☒ p ☐ sf ☐ cf ☐ mf ☐ w *size (μm)*

Processing

primary Powder Metallurgy

specifics

secondary ☐ Casting ☒ Forging ☐ Surface Treatments ☒ Extrusion ☐ Machining ☒ Thermal Treatments

specifics

joining ☐ Brazing ☐ Mechanical ☐ Welding ☐ Other

recyclable ☐ Yes ☐ No

Properties are quoted from manufacturer as general indicators only.

Properties *isotropic* ☐ Yes ☐ No

density ($g\ cm^{-3}$) 2.769

CTE ($x10^6\ K^{-1}$) 19.1

thermal cond. ($Wm^{-1}K^{-1}$) 185

electrical cond. (%IACS)

tensile

yield strength (MPa) 400

ultimate strength (MPa) 455

modulus (GPa) 91.7

elongation (%) 5

Property Advantages ☒ Density ☒ Fatigue ☒ Stiffness ☒ Strength ☐ Thermal Expansion ☐ Thermal Conductivity ☒ Wear ☐ Weldability ☐ Other

Applications structural components with high specific strength and stiffness such as robotics and non-welded tubing assemblies; shafts for sporting equipment; pistons and connecting rods for motorsport; friction components

Products

Sources

Talon Composites, Inc

material (company designation): Talbor® ΔS20 (Delta S20)

Material 6092 / SiC / 20p / T6

matrix Al alloy

reinforcement SiC *volume percent* 20

morphology ☒ p ☐ sf ☐ cf ☐ mf ☐ w *size (μm)*

Processing

primary Powder Metallurgy

specifics

secondary ☐ Casting ☒ Forging ☐ Surface Treatments ☒ Extrusion ☐ Machining ☒ Thermal Treatments

specifics

joining ☐ Brazing ☐ Mechanical ☐ Welding ☐ Other

recyclable ☐ Yes ☐ No

Properties are quoted from manufacturer as general indicators only.

Properties *isotropic* ☐ Yes ☐ No *tensile*

density (g cm^{-3}) 2.796 *yield strength (MPa)* 414

CTE (x10^{6} K^{-1}) 17.5 *ultimate strength (MPa)* 469

thermal cond. (Wm^{-1}K^{-1}) 187 *modulus (GPa)* 101

electrical cond. (%IACS) *elongation (%)* 4.5

Property Advantages ☒ Density ☒ Strength ☒ Wear ☒ Fatigue ☐ Thermal Expansion ☐ Weldability ☒ Stiffness ☐ Thermal Conductivity ☐ Other

Applications structural components with high specific strength and stiffness such as robotics and non-welded tubing assemblies; shafts for sporting equipment; pistons and connecting rods for motorsport; friction components

Products

Sources

Talon Composites, Inc

material (company designation): Talbor® ΔS25 (Delta S25)

Material 6092 / SiC / 25p / T6

matrix Al alloy

reinforcement SiC *volume percent* 25

morphology ☒ p ☐ sf ☐ cf ☐ mf ☐ w *size (μm)*

Processing

primary Powder Metallurgy

specifics

secondary ☐ Casting ☒ Forging ☒ Surface Treatments ☒ Extrusion ☐ Machining ☐ Thermal Treatments

specifics

joining ☐ Brazing ☐ Mechanical ☐ Welding ☐ Other

recyclable ☐ Yes ☐ No

Properties are quoted from manufacturer as general indicators only.

Properties *isotropic* ☐ Yes ☐ No

			tensile
density ($g\ cm^{-3}$)	2.824	*yield strength (MPa)*	434
CTE ($x10^{6}\ K^{-1}$)	16.2	*ultimate strength (MPa)*	490
thermal cond. ($Wm^{-1}K^{-1}$)	187	*modulus (GPa)*	110
electrical cond. (%IACS)		*elongation (%)*	3

Property Advantages ☒ Density ☒ Strength ☒ Wear ☒ Fatigue ☐ Thermal Expansion ☐ Weldability ☒ Stiffness ☐ Thermal Conductivity ☐ Other

Applications structural components with high specific strength and stiffness such as robotics and non-welded tubing assemblies; shafts for sporting equipment; pistons and connecting rods for motorsport; friction components

Products

Sources

Talon Composites, Inc
material (company designation): Talbor® ΔS30 (Delta S30)

Material 6092 / SiC / 30p / T6
matrix Al alloy
reinforcement SiC *volume percent* 30
morphology ☒ p ☐ sf ☐ cf ☐ mf ☐ w *size (μm)*

Processing
primary Powder Metallurgy
specifics

secondary ☐ Casting ☒ Forging ☐ Surface Treatments ☒ Extrusion ☐ Machining ☒ Thermal Treatments
specifics

joining ☐ Brazing ☐ Mechanical ☐ Welding ☐ Other
recyclable ☐ Yes ☐ No

Properties are quoted from manufacturer as general indicators only.

Properties *isotropic* ☐ Yes ☐ No *tensile*
density (g cm^{-3}) 2.849 *yield strength (MPa)* 448
CTE (x10^{6} K^{-1}) 14.9 *ultimate strength (MPa)* 503
thermal cond. (Wm^{-1}K^{-1}) 189 *modulus (GPa)* 120
electrical cond. (%IACS) *elongation (%)* 1.5

Property Advantages ☒ Density ☒ Strength ☒ Wear ☒ Fatigue ☐ Thermal Expansion ☐ Weldability ☒ Stiffness ☐ Thermal Conductivity ☐ Other

Applications structural components with high specific strength and stiffness such as robotics and non-welded tubing assemblies; shafts for sporting equipment; pistons and connecting rods for motorsport; friction components

Products

Sources

Talon Composites, Inc

material (company designation): Talbor® ΔS35 (Delta S35)

Material 6092 / SiC / 35p / T6

matrix Al alloy

reinforcement SiC *volume percent* 35

morphology ☒ p ☐ sf ☐ cf ☐ mf ☐ w *size (μm)*

Processing

primary Powder Metallurgy

specifics

secondary ☐ Casting ☒ Forging ☐ Surface Treatments ☒ Extrusion ☐ Machining ☒ Thermal Treatments

specifics

joining ☐ Brazing ☐ Mechanical ☐ Welding ☐ Other

recyclable ☐ Yes ☐ No

Properties are quoted from manufacturer as general indicators only.

Properties *isotropic* ☐ Yes ☐ No

density (g cm^{-3}) 2.876

CTE (x10^{6} K^{-1}) 14.2

thermal cond. (Wm^{-1}K^{-1}) 189

electrical cond. (%IACS)

tensile

yield strength (MPa) 462

ultimate strength (MPa) 524

modulus (GPa) 130

elongation (%) 1.0

Property Advantages ☒ Density ☒ Fatigue ☒ Stiffness ☒ Strength ☐ Thermal Expansion ☐ Thermal Conductivity ☒ Wear ☐ Weldability ☐ Other

Applications structural components with high specific strength and stiffness such as robotics and non-welded tubing assemblies; shafts for sporting equipment; pistons and connecting rods for motorsport; friction components

Products

Sources

Talon Composites, Inc

material (company designation): Talbor® ΔS40 (Delta S40)

Material 6092 / SiC / 40p / T6

matrix Al alloy

reinforcement SiC *volume percent* 40

morphology ☒ p ☐ sf ☐ cf ☐ mf ☐ w *size (μm)*

Processing

primary Powder Metallurgy

specifics

secondary ☐ Casting ☒ Forging ☐ Surface Treatments ☒ Extrusion ☐ Machining ☒ Thermal Treatments

specifics

joining ☐ Brazing ☐ Mechanical ☐ Welding ☐ Other

recyclable ☐ Yes ☐ No

Properties are quoted from manufacturer as general indicators only.

Properties *isotropic* ☐ Yes ☐ No *tensile*

density ($g\ cm^{-3}$) 2.904 *yield strength (MPa)* 469

CTE ($x10^6\ K^{-1}$) 13.7 *ultimate strength (MPa)* 538

thermal cond. ($Wm^{-1}K^{-1}$) 190 *modulus (GPa)* 142

electrical cond. (%IACS) *elongation (%)* 0.8

Property Advantages ☒ Density ☒ Strength ☒ Wear ☒ Fatigue ☐ Thermal Expansion ☐ Weldability ☒ Stiffness ☐ Thermal Conductivity ☐ Other

Applications structural components with high specific strength and stiffness such as robotics and non-welded tubing assemblies; shafts for sporting equipment; pistons and connecting rods for motorsport; friction components

Products

Sources

Talon Composites, Inc

material (company designation): Talbor® ΩB5 (Omega B5)

Material 7093 / B4C / 5p / T6

matrix Al alloy

reinforcement B4C *volume percent* 5

morphology ☒ p ☐ sf ☐ cf ☐ mf ☐ w *size (μm)*

Processing

primary Powder Metallurgy

specifics

secondary ☐ Casting ☒ Forging ☐ Surface Treatments ☒ Extrusion ☐ Machining ☒ Thermal Treatments

specifics

joining ☐ Brazing ☐ Mechanical ☐ Welding ☐ Other

recyclable ☐ Yes ☐ No

Properties are quoted from manufacturer as general indicators only.

Properties *isotropic* ☐ Yes ☐ No

			tensile
density (g cm⁻³)	2.849	*yield strength (MPa)*	586
CTE (x10⁶ K⁻¹)	22.5	*ultimate strength (MPa)*	621
thermal cond. (Wm⁻¹K⁻¹)	128	*modulus (GPa)*	77.9
electrical cond. (%IACS)		*elongation (%)*	3

Property Advantages ☒ Density ☒ Strength ☒ Wear ☒ Fatigue ☐ Thermal Expansion ☒ Weldability ☒ Stiffness ☐ Thermal Conductivity ☐ Other

Applications for high strength and high stiffness applications

Products

Sources

Talon Composites, Inc

material (company designation): Talbor® ΩB10 (Omega B10)

Material 7093 / B4C / 10p / T6

matrix Al alloy

reinforcement B4C *volume percent* 10

morphology ☒ p ☐ sf ☐ cf ☐ mf ☐ w *size (μm)*

Processing

primary Powder Metallurgy

specifics

secondary ☐ Casting ☒ Forging ☐ Surface Treatments ☒ Extrusion ☐ Machining ☒ Thermal Treatments

specifics

joining ☐ Brazing ☐ Mechanical ☐ Welding ☐ Other

recyclable ☐ Yes ☐ No

Properties are quoted from manufacturer as general indicators only.

Properties *isotropic* ☐ Yes ☐ No

			tensile
density (g cm^{-3})	2.835	*yield strength (MPa)*	586
CTE (x10^{6} K^{-1})	21.1	*ultimate strength (MPa)*	621
thermal cond. (Wm^{-1}K^{-1})	121	*modulus (GPa)*	86.2
electrical cond. (%IACS)		*elongation (%)*	2

Property Advantages ☒ Density ☒ Strength ☒ Wear ☒ Fatigue ☐ Thermal Expansion ☒ Weldability ☒ Stiffness ☐ Thermal Conductivity ☐ Other

Applications for high strength and high stiffness applications

Products

Sources

Talon Composites, Inc

material (company designation): Talbor® ΩB15 (Omega B15)

Material 7093 / B4C / 15p / T6

matrix Al alloy

reinforcement B4C *volume percent* 15

morphology ☒ p ☐ sf ☐ cf ☐ mf ☐ w *size (μm)*

Processing

primary Powder Metallurgy

specifics

secondary ☐ Casting ☒ Forging ☐ Surface Treatments ☒ Extrusion ☐ Machining ☒ Thermal Treatments

specifics

joining ☐ Brazing ☐ Mechanical ☐ Welding ☐ Other

recyclable ☐ Yes ☐ No

Properties are quoted from manufacturer as general indicators only.

Properties *isotropic* ☐ Yes ☐ No *tensile*

density ($g\ cm^{-3}$) 2.818 *yield strength (MPa)* 586

CTE ($x10^6\ K^{-1}$) 19.1 *ultimate strength (MPa)* 621

thermal cond. ($Wm^{-1}K^{-1}$) 114 *modulus (GPa)* 93.8

electrical cond. (%IACS) *elongation (%)* 1

Property Advantages ☒ Density ☒ Strength ☒ Wear ☒ Fatigue ☐ Thermal Expansion ☒ Weldability ☒ Stiffness ☐ Thermal Conductivity ☐ Other

Applications for high strength and high stiffness applications

Products

Sources

Talon Composites, Inc

material (company designation): Talbor® ΩS5 (Omega S5)

Material 7093 / SiC / 5p / T6

matrix Al alloy

reinforcement SiC *volume percent* 5

morphology ☒ p ☐ sf ☐ cf ☐ mf ☐ w *size (μm)*

Processing

primary Powder Metallurgy

specifics

secondary ☐ Casting ☒ Forging ☐ Surface Treatments ☒ Extrusion ☐ Machining ☒ Thermal Treatments

specifics

joining ☐ Brazing ☐ Mechanical ☐ Welding ☐ Other

recyclable ☐ Yes ☐ No

Properties are quoted from manufacturer as general indicators only.

Properties *isotropic* ☐ Yes ☐ No

density ($g\ cm^{-3}$)	2.885	*tensile yield strength (MPa)*	607
CTE ($x10^6\ K^{-1}$)	22.5	*ultimate strength (MPa)*	648
thermal cond. ($Wm^{-1}K^{-1}$)	131	*modulus (GPa)*	77.9
electrical cond. (%IACS)		*elongation (%)*	4

Property Advantages ☒ Density ☒ Strength ☒ Wear ☒ Fatigue ☐ Thermal Expansion ☐ Weldability ☒ Stiffness ☐ Thermal Conductivity ☐ Other

Applications structural components with high specific strength and stiffness such as robotics; shafts for sporting equipment; pistons for motorsport; skid plates

Products

Sources

Talon Composites, Inc

material (company designation): Talbor® ΩS10 (Omega S10)

Material 7093 / SiC / 10p / T6

matrix Al alloy

reinforcement SiC *volume percent* 10

morphology ☒ p ☐ sf ☐ cf ☐ mf ☐ w *size (μm)*

Processing

primary Powder Metallurgy

specifics

secondary ☐ Casting ☐ Forging ☐ Surface Treatments ☐ Extrusion ☐ Machining ☐ Thermal Treatments

specifics

joining ☐ Brazing ☐ Mechanical ☐ Welding ☐ Other

recyclable ☐ Yes ☐ No

Properties are quoted from manufacturer as general indicators only.

Properties *isotropic* ☐ Yes ☐ No *tensile*

density (g cm^{-3}) 2.901 *yield strength (MPa)* 614

CTE (x10^{6} K^{-1}) 21.1 *ultimate strength (MPa)* 655

thermal cond. (Wm^{-1}K^{-1}) 134 *modulus (GPa)* 85.5

electrical cond. (%IACS) *elongation (%)* 3

Property Advantages ☒ Density ☒ Strength ☒ Wear ☒ Fatigue ☐ Thermal Expansion ☐ Weldability ☒ Stiffness ☐ Thermal Conductivity ☐ Other

Applications structural components with high specific strength and stiffness such as robotics; shafts for sporting equipment; pistons for motorsport; skid plates

Products

Sources

Talon Composites, Inc

material (company designation): Talbor® ΩS15 (Omega S15)

Material 7093 / SiC / 15p / T6

matrix Al alloy

reinforcement SiC *volume percent* 15

morphology ☒ p ☐ sf ☐ cf ☐ mf ☐ w *size (μm)*

Processing

primary Powder Metallurgy

specifics

secondary ☐ Casting ☐ Forging ☐ Surface Treatments ☐ Extrusion ☐ Machining ☐ Thermal Treatments

specifics

joining ☐ Brazing ☐ Mechanical ☐ Welding ☐ Other

recyclable ☐ Yes ☐ No

Properties are quoted from manufacturer as general indicators only.

Properties *isotropic* ☐ Yes ☐ No *tensile*

density (g cm^{-3}) 2.920 *yield strength (MPa)* 621

CTE (x10^{6} K^{-1}) 19.1 *ultimate strength (MPa)* 662

thermal cond. (Wm^{-1}K^{-1}) 137 *modulus (GPa)* 93.8

electrical cond. (%IACS) *elongation (%)* 1.5

Property Advantages ☒ Density ☒ Strength ☒ Wear ☒ Fatigue ☐ Thermal Expansion ☐ Weldability ☒ Stiffness ☐ Thermal Conductivity ☐ Other

Applications structural components with high specific strength and stiffness such as robotics; shafts for sporting equipment; pistons for motorsport; skid plates

Products

Sources

TCI Automotive

151 Industrial Drive

Ashland MS 38603 USA

tel ++ 1 (662) 224 8972
fax ++ 1 (662) 224 8255
email
web address www.tciauto.com

TCI Automotive manufactures two metal matrix composite trans brake release valves for its transmission valve bodies. The valve body of the transmission is referred to as being the central nervous system that controls operation of the transmission since it controls the hydraulic pressure that directs the reaction members that release and apply the bands and clutches. The valve body, usually of aluminium or iron-based materials, has bores and fluid passages machined out of it. These bores are fitted with various valves, one being the release valve for which TCI offers a version from MMC material.

Technical Dynamics Aluminum

9124 S.E. 64th Avenue

Portland OR 97206 USA

tel ++ 1 (503) 775 6791 // 1 (800) 677 6791
fax ++ 1 (503) 775 1614
email sdemas@tda-ano.com
web address www.tda-ano.com

Technical Dynamics Aluminum specialises in the extrusion of aluminium-based products, including MMC. The company appears focussed on the future success of MMC citing the economic benefits of aluminium-based MMC compared to competing materials such as titanium alloys. Technical Dynamics Aluminum offers 3 inch circle size extrusions for MMC materials and certifies to all US military and federal specifications, including quality standard MIL-1-45208A.

Thermal Ceramics
2102 Old Savannah Road
Augusta GA 30903 USA
tel ++ 1 (706) 796 4200
fax ++ 1 (706) 796 4398
email tceramics@thermalceramics.com
web address www.thermalceramics.com

Thermal Ceramics manufactures precision-shaped ceramic fibre preforms suitable for infiltration with molten metal by processes such as squeeze casting. These preforms, called Fiberalloy™, range from discontinuous alumina-silicate (Kaowool®) fibres and high-purity alumina (Saffil®) fibres to ceramic powders. The preforms are fabricated from wet slurry of cleaned, composite-grade ceramic fibre and binder. The slurry undergoes a vacuum forming process that orientates the fibres into a two-dimensional random planar mat with at least 70% of the fibres parallel to the forming surface. Preforms of square or rectangular cross-sections are most economical; however, other geometries can be produced.

Tikka-Nastat Oy

Kirkkokatu 11

Tikkakoski 41160 Finland

tel ++ 358 (14) 375 9200
fax ++ 358 (14) 375 9220
email
web address www.tikka-nastat.fi

Due to government legislation in Finland, steel tyre studs were banned as they excessively damaged the road surface. In response Tikka-Nastat (and reportedly several other manufacturers) developed tyre studs with a particle-reinforced aluminium body. The Kometa Marathon tyre stud manufactured by Tikka-Nastat includes a cemented carbide pin in the centre of a particle-reinforced aluminium body: 6061/Al2O3/15p, hot-rolled from the billet, drawn into wire and then formed into the body. These tyre studs are specifically designed for Nordic conditions.

Toyota Motor Corporation
1, Toyota-cho
Aichi 471-8571 Japan
tel ++ 81 (565) 28 2121
fax ++ 81 (565) 23 5800
email
web address http://www.toyota.com/

Toyota has pioneered the integration of MMC technology into the automotive industry. In the early 1980s, Toyota introduced its first reinforced-metal component, a diesel piston with a selectively reinforced ring groove, produced by its ArtMetal subsidiary. Today, Toyota applies MMC material systems in a variety of automotive applications in a range of vehicles. The 2ZZ-GE engine for the Toyota Celica has MMC cylinder bores: the all-aluminium cylinder block is reinforced in the bore with 5 vol. % alumina-silica short fibres and 10 vol. % mullite particles. The application of this material system allowed the bore-to-bore distance to be decreased from 8.5 mm to 5.5 mm, thus allowing the bore size to increase, and shortening the piston stroke. The net effect of these alterations is higher engine performance. In recent years, the application of MMC systems at Toyota has reached beyond engine components; for example, front brake discs for the RAV4-EV electric vehicle are cast from silicon carbide particulate-reinforced aluminium. In addition, Al-SiC materials from dmc2 Electronic Components (DEC) are used for thermal management in IGBT (Insulated Gate Bipolar Transistor) power modules for the Prius hybrid vehicle.

Triton Systems Inc

200 Turnpike Road

Chelmsford MA 01824 USA

tel ++ 1 (978) 250 4200
fax ++ 1 (978) 250 4533
email tsi-info@tritonsys.com
web address www.tritonsys.com

Triton Systems is a materials company with a diverse product portfolio including MMC materials. Its MMC activities encompass the electronic packaging market in addition to aerospace and military applications. The Enhanced Pressure Infiltration Casting (EPIC™) process, used by Triton to produce aluminium-matrix composites, is based on conventional pressure infiltration processes, but makes use of innovative mould tooling and preform fabrication techniques to enable rapid prototyping and economic low volume production. EPIC™ is used to manufacture two MMC systems: HiPAC™ continuous fibre reinforced aluminium, and HiPAX™ discontinuous silicon carbide particle reinforced aluminium.

HiPAC™ is produced by the infiltration of fibre preforms, prepared by either filament winding or braiding, to produce a composite exhibiting high strength and excellent fatigue behaviour. These properties make the material suitable for high performance actuator housings and elevated-temperature pressure vessels. Hamilton Standards' advanced lightweight nozzle actuator for Pratt and Whitney's convergent nozzle flap is an application that replaces a more expensive and heavier titanium component.

HiPAX™ is produced near-net shape, and the concentration of ceramic can be controlled to optimise performance for specific applications. These materials are suited for gas turbine components such as stator vanes, hydraulic manifolds, and opto-mechanical assemblies such as gimbal rings and optical mounts.

Additionally, Triton's Laser Free Form Fabrication (LF3™) process clads metal powder (titanium-, copper-, magnesium-based) with reinforcement powder (TiC, B_4C, SiC). The process is computer-controlled, such that it is amenable to solid freeform fabrication. The company offers a titanium-based material, referred to as TiLAS™ and described as "a modified titanium alloy in which very strong reinforcement phases precipitate during laser processing."

Triton Systems Inc

material (company designation): HiPAX™

Material	A206 / SiC / 40p		
matrix	Al alloy		
reinforcement	SiC	*volume percent*	40
morphology	☒ p ☐ sf ☐ cf ☐ mf ☐ w	*size (μm)*	

Processing	
primary	Infiltration
specifics	Enhanced Pressure Infiltration Casting, EPIC™
secondary	☐ Casting ☐ Forging ☐ Surface Treatments ☐ Extrusion ☐ Machining ☐ Thermal Treatments
specifics	
joining	☐ Brazing ☐ Mechanical ☐ Welding ☐ Other
recyclable	☐ Yes ☐ No

Properties are quoted from manufacturer as general indicators only.

Properties	*isotropic* ☐ Yes ☐ No		*tensile*
density (g cm⁻³)		*yield strength (MPa)*	
CTE (x10⁶ K⁻¹)		*ultimate strength (MPa)*	
thermal cond. (Wm⁻¹K⁻¹)		*modulus (GPa)*	
electrical cond. (%IACS)		*elongation (%)*	

Property Advantages	☒ Density	☒ Strength	☐ Wear
	☐ Fatigue	☒ Thermal Expansion	☐ Weldability
	☒ Stiffness	☐ Thermal Conductivity	☐ Other

Applications	structural components such as gimbal rings and optical mounts (opto-mechanical assembly components), hydraulic manifolds, stator vanes (gas turbines)
Products	V-22 Osprey tiltrotor hydraulic manifold
Sources	

Twin Fin Composites

PO Box 185
2844 Industrial Park Drive
Austinburg OH 44010 USA

tel ++ 1 (440) 275 2255
fax ++ 1 (440) 275 1255
email info@twinfin.com
web address www.twinfin.com

Twin Fin has been supplying air-cooled, MMC brakes for the converting industry for more than a decade. They currently offer brake discs from 10 to 30 inches in diameter for nontransportation applications, corresponding to loading from 125 to 57,000 in-ibs. The brake discs are manufactured from silicon carbide-reinforced aluminium, offering significant benefits compared to conventional cast iron. The MMC exhibits higher heat transfer rates, improving heat dissipation and thus reducing the overall size and cost of the brake. In addition, the heat dissipation characteristics allow the friction material to remain in phase and smear on the discs, creating a hard surface. The result is improved lifetime of both the friction material and the discs. The usage of an aluminium matrix gives a lower inertia compared to cast iron, thus improving the tension control response. These brake discs have been used successfully by over 30 companies.

TWM Technology Ltd

Sheerspeed House
Heathpark Way
Honiton, Devon EX14 8BB UK

tel ++ 44 (1404) 46737
fax ++
email TWMTech@eclipse.co.uk
web address home1.gte.net/techexp/twm.htm

TWM Technology manufactures metal packages and enclosures for the microelectronics industry. TWM's product range includes plug-ins, TO-series, flatpack packages, fibre-optic, microwave and power packages and encapsulations. TWM has the capability to offer these packaging solutions from MMC, in particular from the Al-SiC family. They also manufacture these components from so called "Al-Si in-situ MMC" produced by Osprey Metals, which is in all likelihood (according to the definition of MMC used within this text) a rapidly solidified alloy and not MMC. These packages are broadly aimed at microelectronics, including military, aerospace, industrial, commercial, telecommunications, and automotive. In addition to producing MMC components, TWM has the capability to plate and machine in-house, giving it complete control of the manufacturing process, including rapid prototyping.

In the US:
tel++ 1(760) 360 8465 or 1 (760) 360 4503
email: techexp@gte.net

U.S. Chrome Corporation

650 Oak Park Avenue

Fond du Lac WI 54935 USA

tel ++ 1 (716) 343 7077
fax ++ 1 (716) 344 4016
email nicom@thesurf.com
web address www.uschrome.com (www.usnicom.com)

U.S. Chrome is an established manufacturer of coatings for a variety of applications including combustion engines. The deposition process of their coatings is proprietary, but they have developed and successfully marketed MMC coatings based on nickel reinforced with silicon carbide. This coating is called NiCom®, and is electroplated onto substrate materials. The resulting microstructure is comprised of a nickel matrix with a uniform distribution of silicon carbide particles having an average diameter of 3 μm.

The coating is marketed as both a method of repair for used cylinders and as a way to enhance the performance of new engines. With regard to repair, such coatings are suitable for restoring worn, gouged and scored cylinder bores. The coatings are said to exhibit excellent adhesion to the substrate material, although tensile and shear strength data were not provided. The material exhibits minimum hardness of 550 Vickers (53 Rockwell C), improved wear resistance compared to chromium, and improved corrosion resistance compared to electrolytic nickel. Frictional properties are also good due to the oleophilic nature of the coating.

NiCom® coatings have successfully been applied in a variety of internal combustion engines: automobile, aircraft, marine, snowmobile, and motorbike. Other applications of nickel carbide coatings include oil and gas well tubes. U.S. Chrome is an official sponsor of Team Kawasaki providing cylinder coatings for their motorcycle racing team.

Unifrax Corporation

2351 Whirlpool Street

Niagara Falls NY 14305-2413 USA

tel ++ 1 (716) 278 3800
fax ++ 1 (716) 278 3900
email
web address www.unifrax.com

Unifrax manufactures preforms for infiltration to produce metal matrix composites (in addition to its core activities in fibrous ceramic materials). The Prefrax® family of preforms are formed from aluminosilicate fibres. These fibres are manufactured by the initial melting of a mix of alumina and high purity silica at 1926°C. The molten material is subsequently sheared by a molten stream of the same material, which shapes it into fibres. A binder is added and the mixture formed into near net shapes. The water is removed, then the part is calcinated and finished into the preform. Unifrax also produces Premax™ preforms, which are formed from discontinuous fibres composed of nominally alumina. Both families of preform contain 10 to 20% fibres by volume and can be machined to tight tolerances.

These preforms can be pressure infiltrated with metals such as aluminium to produce short fibre reinforced metals (SFRM). For example, squeeze casting can be used to produce reinforced-aluminium pistons that feature reduced piston wear in-service and improved high-temperature strength.

Visteon Corporation

5500 Auto Club Drive

Dearborn MI 48126 USA

tel ++
fax ++
email
web address www.visteon.com

Visteon is a supplier to the automotive industry. Its MMC activity is focussed on driveshafts and these are produced for Ford's Crown Victoria Police Interceptor (automobile). The driveshaft is manufactured from seamless MMC tube extruded by Alcoa from Duralcan material (Alcan). The MMC driveshafts are available in four diameters, hence offering a range of strengths.

Volkswagen

Germany

tel ++
fax ++
email
web address

Volkswagen has successfully implemented MMC technology into production vehicles. The rear brake drums on the compact Lupo 3L model are fabricated from a silicon carbide-reinforced aluminium alloy. The material A359/SiC/20p is produced as ingots by Alcan then shipped to Kolbenschmidt Pierburg in Germany for casting. The cast discs are machined to their final form by Volkswagen. These discs may become an integral part of an axle assembly that will be applied in the Audi A2 model.

Walt Disney

USA

tel ++
fax ++
email
web address

DWA Aluminum Composites has produced brake fins (6092/SiC/17.5p) for the Big Thunder Railway thrill ride at Walt Disney World, demonstrating performance in a critical application.

Western Motorsports

no. 4
1115 - 48 Avenue SE
Calgary, Alberta T2G 2A7 Canada

tel ++	1 (403) 243 6205
fax ++	1 (403) 243 8102
email	fastford@wmsracing.com
web address	www.wmsracing.com

Western motorsport produces the Cobra, a SuperComp sports car, with MMC vented and cross-drilled brake discs. These discs are half the weight of conventional brake discs, giving lower unsprung mass and thus providing better acceleration and handling.

z- Aavid Thermal Technologies Incorporated

Suite 509
One Eagle Square
Concord NH 03301 USA

tel ++ 1 (603) 224 1117
fax ++ 1 (603) 224 6673
email
web address www.aavid.com

This company is no longer active in the MMC industry, as indicated by the "z-" preceding its name.

Aavid is a producer of advanced thermal management products, some of which made use of MMC materials. Aavid does not produce MMC, but purchased Al-SiC materials from Ceramic Processing Systems (CPS).

z- Advanced Composite Materials Corporation (ACMC)

1525 South Buncombe Road

Greer SC 29651 USA

tel ++ 1 (864) 877 0123
fax ++ 1 (864) 879 6615
email
web address

This company is no longer active in the MMC industry, as indicated by the "z-" preceding its name.

Advanced Composite Materials Corporation (ACMC) developed a family of high-performance silicon carbide-reinforced aluminium materials for military applications, including an optical grade (2009/SiC/30p), an instrument/guidance grade (6061/SiC/40p), and a structural grade reinforced with whiskers (2009/SiC/15w). ACMC has ceased production of MMC materials, concentrating on the production of SiC whiskers.

ACMC licences its MMC technology; for further information contact Paul Hood, Vice President, Sales at ACMC.

z- Advanced Materials Lanxide (AML)

High Performance Materials Group (HPMG), MSE, Inc.
1300 Marrows Road
Newark DE 19711 USA

tel ++ 1 (302) 456 6200
fax ++ 1 (302) 454 1712
email
web address www.lanxide.com

This company no longer exists, as indicated by the "z-" preceding its name.

Advanced Materials Lanxide is now High Performance Materials Group owned solely by MSE, Inc.

Virgil Irick VP Technology
virick@dpnet.net
++1 (302) 456 6543

z- Alyn Corporation

USA

tel ++
fax ++
email
web address

This company no longer exists, thus it is not active in the MMC industry, as indicated by the "z-" preceding its name.

Alyn Corporation filed for bankruptcy and closed its doors in July 1999. Pechiney Japon was the exclusive agent of Alyn in Japan.

Alyn manufactured particle-reinforced aluminium materials by a powder metallurgical route; reinforcements included both silicon carbide and boron carbide. These Boralyn materials were produced by cold isostatically pressing powders into a green form, followed by sintering and extrusion to produce the final microstructure.

Alyn supplied material for a diverse range of applications such as bicycle frames for the Ikon/Lexus racing team, heads for golf clubs, casing for a watch from Reebok, and the engine cradle for the EV-1 from General Motors. A number of other development efforts were also reported: snowshoes, nuclear shielding applications in spent fuel dry storage by Transnuclear, engine blocks in two-stroke combustion engines, and applications on the Cessna aircraft. Alyn also pursued application of Boralyn for substrates of computer hard disks, and it was found that its MMC outperformed competing materials such as aluminium, glass and a glass ceramic.

Previous contact information:

Alyn Corporation
16761 Hale Avenue
Irvine CA 92606
tel ++1 (949) 475 1525, fax ++1 (949) 475 2359
engineerinfo@alyn.com
www.alyn.com

z- Alyn Corporation

material (company designation): Boralyn HS15 - T6

Material 6092 / SiC / 15p / extruded / T6

matrix Al alloy

reinforcement SiC *volume percent* 15

morphology ☒ p ☐ sf ☐ cf ☐ mf ☐ w *size (μm)*

Processing

primary Powder Metallurgy

specifics

secondary ☒ Casting ☒ Forging ☐ Surface Treatments ☒ Extrusion ☒ Machining ☐ Thermal Treatments

specifics

joining ☐ Brazing ☐ Mechanical ☒ Welding ☐ Other

recyclable ☐ Yes ☐ No

Properties are quoted from manufacturer as general indicators only.

Properties *isotropic* ☐ Yes ☒ No *tensile*

density (g cm⁻³) 2.79 — *yield strength (MPa)* 430

CTE (x10⁶ K⁻¹) 18.9 — *ultimate strength (MPa)* 500

thermal cond. (Wm⁻¹K⁻¹) — *modulus (GPa)* 98

electrical cond. (%IACS) — *elongation (%)* 7.0

Property Advantages ☒ Density ☒ Strength ☒ Wear ☒ Fatigue ☐ Thermal Expansion ☐ Weldability ☒ Stiffness ☐ Thermal Conductivity ☐ Other

Applications

Products Flash 300 Boralyn watch with MMC case

Sources Froes, F.H. (1999).

z- Alyn Corporation

material (company designation): Boralyn H25 - T6

Material 6092 / B4C / 25p / extruded / T6

matrix Al alloy

reinforcement B4C *volume percent* 25

morphology ☒ p ☐ sf ☐ cf ☐ mf ☐ w *size (μm)*

Processing

primary Powder Metallurgy

specifics

secondary ☒ Casting ☒ Forging ☐ Surface Treatments ☒ Extrusion ☒ Machining ☐ Thermal Treatments

specifics

joining ☐ Brazing ☐ Mechanical ☒ Welding ☐ Other

recyclable ☐ Yes ☐ No

Properties are quoted from manufacturer as general indicators only.

Properties *isotropic* ☐ Yes ☒ No *tensile*

density ($g\ cm^{-3}$) 2.66 *yield strength (MPa)* 390

CTE ($x10^6\ K^{-1}$) 16.2 *ultimate strength (MPa)* 480

thermal cond. ($Wm^{-1}K^{-1}$) *modulus (GPa)* 120

electrical cond. (%IACS) *elongation (%)* 3.5

Property Advantages ☒ Density ☒ Strength ☒ Wear ☒ Fatigue ☐ Thermal Expansion ☒ Weldability ☒ Stiffness ☐ Thermal Conductivity ☐ Other

Applications

Products

Sources Froes, F.H. (1999).

z- Alyn Corporation

material (company designation): Boralyn E5 - T6

Material 7093 / B4C / 5p / extruded / T6

matrix Al alloy

reinforcement B4C *volume percent* 5

morphology ☒ p ☐ sf ☐ cf ☐ mf ☐ w *size (μm)*

Processing

primary Powder Metallurgy

specifics

secondary ☒ Casting ☒ Forging ☐ Surface Treatments ☒ Extrusion ☒ Machining ☐ Thermal Treatments

specifics

joining ☐ Brazing ☐ Mechanical ☒ Welding ☐ Other

recyclable ☐ Yes ☐ No

Properties are quoted from manufacturer as general indicators only.

Properties *isotropic* ☐ Yes ☒ No

			tensile
density (g cm^{-3})		*yield strength (MPa)*	590
CTE (x10^{6} K^{-1})	20.52	*ultimate strength (MPa)*	660
thermal cond. (Wm^{-1}K^{-1})		*modulus (GPa)*	79
electrical cond. (%IACS)		*elongation (%)*	8.5

Property Advantages ☒ Density ☒ Fatigue ☒ Stiffness ☒ Strength ☐ Thermal Expansion ☐ Thermal Conductivity ☒ Wear ☒ Weldability ☐ Other

Applications

Products

Sources Froes, F.H. (1999).

z- Alyn Corporation

material (company designation): Boralyn H15 - T6

Material 6092 / B4C / 15p / extruded / T6

matrix Al alloy

reinforcement B4C *volume percent* 15

morphology ☒ p ☐ sf ☐ cf ☐ mf ☐ w *size (μm)*

Processing

primary Powder Metallurgy

specifics

secondary ☒ Casting ☒ Forging ☐ Surface Treatments ☒ Extrusion ☒ Machining ☒ Thermal Treatments

specifics

joining ☐ Brazing ☐ Mechanical ☒ Welding ☐ Other

recyclable ☐ Yes ☐ No

Properties are quoted from manufacturer as general indicators only.

Properties *isotropic* ☐ Yes ☒ No

density ($g\ cm^{-3}$)	2.68	*tensile yield strength (MPa)*	370
CTE ($x10^6\ K^{-1}$)	18.9	*ultimate strength (MPa)*	450
thermal cond. ($Wm^{-1}K^{-1}$)		*modulus (GPa)*	98
electrical cond. (%IACS)		*elongation (%)*	5.8

Property Advantages ☒ Density ☒ Strength ☒ Wear ☒ Fatigue ☐ Thermal Expansion ☒ Weldability ☒ Stiffness ☐ Thermal Conductivity ☐ Other

Applications

Products

Sources Froes, F.H. (1999).

z- Alyn Corporation

material (company designation): Boralyn HS25 - T6

Material 6092 / SiC / 25p / extruded / T6

matrix Al alloy

reinforcement SiC *volume percent* 25

morphology ☒ p ☐ sf ☐ cf ☐ mf ☐ w *size (μm)*

Processing

primary Powder Metallurgy

specifics

secondary ☒ Casting ☒ Forging ☐ Surface Treatments ☒ Extrusion ☒ Machining ☒ Thermal Treatments

specifics

joining ☐ Brazing ☐ Mechanical ☒ Welding ☐ Other

recyclable ☐ Yes ☐ No

Properties are quoted from manufacturer as general indicators only.

Properties *isotropic* ☐ Yes ☒ No *tensile*

density (g cm^{-3}) 2.84 *yield strength (MPa)* 460

CTE (x10^{6} K^{-1}) 15.3 *ultimate strength (MPa)* 540

thermal cond. (Wm^{-1}K^{-1}) *modulus (GPa)* 120

electrical cond. (%IACS) *elongation (%)* 4.5

Property Advantages ☒ Density ☒ Fatigue ☒ Stiffness ☒ Strength ☐ Thermal Expansion ☐ Thermal Conductivity ☒ Wear ☐ Weldability ☐ Other

Applications

Products

Sources Froes, F.H. (1999).

z- Alyn Corporation

material (company designation): Boralyn H35 - T6

Material 6092 / B4C / 35p / extruded / T6

matrix Al alloy

reinforcement B4C *volume percent* 35

morphology ☒ p ☐ sf ☐ cf ☐ mf ☐ w *size (μm)*

Processing

primary Powder Metallurgy

specifics

secondary ☒ Casting ☒ Forging ☐ Surface Treatments ☒ Extrusion ☒ Machining ☒ Thermal Treatments

specifics

joining ☐ Brazing ☐ Mechanical ☒ Welding ☐ Other

recyclable ☐ Yes ☐ No

Properties are quoted from manufacturer as general indicators only.

Properties *isotropic* ☐ Yes ☒ No

			tensile
density ($g\ cm^{-3}$)	2.66	*yield strength (MPa)*	450
CTE ($x10^6\ K^{-1}$)	14.58	*ultimate strength (MPa)*	520
thermal cond. ($Wm^{-1}K^{-1}$)		*modulus (GPa)*	130
electrical cond. (%IACS)		*elongation (%)*	1.0

Property Advantages ☒ Density ☒ Strength ☒ Wear ☒ Fatigue ☐ Thermal Expansion ☒ Weldability ☒ Stiffness ☐ Thermal Conductivity ☐ Other

Applications

Products

Sources Froes, F.H. (1999).

z- Alyn Corporation

material (company designation): Boralyn HS35 - T6

Material 6092 / SiC / 35p / extruded / T6

matrix Al alloy

reinforcement SiC *volume percent* 35

morphology ☒ p ☐ sf ☐ cf ☐ mf ☐ w *size (μm)*

Processing

primary Powder Metallurgy

specifics

secondary ☒ Casting ☒ Forging ☐ Surface Treatments ☒ Extrusion ☒ Machining ☒ Thermal Treatments

specifics

joining ☐ Brazing ☐ Mechanical ☒ Welding ☐ Other

recyclable ☐ Yes ☐ No

Properties are quoted from manufacturer as general indicators only.

Properties *isotropic* ☐ Yes ☒ No *tensile*

density ($g\ cm^{-3}$) 2.90 *yield strength (MPa)* 520

CTE ($x10^6\ K^{-1}$) 14.22 *ultimate strength (MPa)* 590

thermal cond. ($Wm^{-1}K^{-1}$) *modulus (GPa)* 130

electrical cond. (%IACS) *elongation (%)* 1.2

Property Advantages ☒ Density ☒ Strength ☒ Wear ☒ Fatigue ☐ Thermal Expansion ☐ Weldability ☒ Stiffness ☐ Thermal Conductivity ☐ Other

Applications

Products

Sources Froes, F.H. (1999).

z- Alyn Corporation

material (company designation): Boralyn E10 - T6

Material 7093 / B4C / 10p / extruded / T6

matrix Al alloy

reinforcement B4C *volume percent* 10

morphology ☒ p ☐ sf ☐ cf ☐ mf ☐ w *size (μm)*

Processing

primary Powder Metallurgy

specifics

secondary ☒ Casting ☒ Forging ☐ Surface Treatments ☒ Extrusion ☒ Machining ☐ Thermal Treatments

specifics

joining ☐ Brazing ☐ Mechanical ☒ Welding ☐ Other

recyclable ☐ Yes ☐ No

Properties are quoted from manufacturer as general indicators only.

Properties *isotropic* ☐ Yes ☒ No

			tensile
density ($g\ cm^{-3}$)	2.74	*yield strength (MPa)*	610
CTE ($x10^6\ K^{-1}$)	19.8	*ultimate strength (MPa)*	660
thermal cond. ($Wm^{-1}K^{-1}$)		*modulus (GPa)*	91
electrical cond. (%IACS)		*elongation (%)*	7.9

Property Advantages ☒ Density ☒ Strength ☒ Wear ☒ Fatigue ☐ Thermal Expansion ☒ Weldability ☒ Stiffness ☐ Thermal Conductivity ☐ Other

Applications

Products

Sources Froes, F.H. (1999).

z- Alyn Corporation

material (company designation): Boralyn E15 - T6

Material 7093 / B4C / 15p / extruded / T6

matrix Al alloy

reinforcement B4C *volume percent* 15

morphology ☒ p ☐ sf ☐ cf ☐ mf ☐ w *size (μm)*

Processing

primary Powder Metallurgy

specifics

secondary ☒ Casting ☒ Forging ☐ Surface Treatments ☒ Extrusion ☒ Machining ☐ Thermal Treatments

specifics

joining ☐ Brazing ☐ Mechanical ☒ Welding ☐ Other

recyclable ☐ Yes ☐ No

Properties are quoted from manufacturer as general indicators only.

Properties *isotropic* ☐ Yes ☒ No

density ($g\ cm^{-3}$)	2.74	*tensile yield strength (MPa)*	620
CTE ($x10^{6}\ K^{-1}$)	18.9	*ultimate strength (MPa)*	680
thermal cond. ($Wm^{-1}K^{-1}$)		*modulus (GPa)*	100
electrical cond. (%IACS)		*elongation (%)*	7.9

Property Advantages ☒ Density ☒ Strength ☒ Wear ☒ Fatigue ☐ Thermal Expansion ☒ Weldability ☒ Stiffness ☐ Thermal Conductivity ☐ Other

Applications

Products

Sources Froes, F.H. (1999).

z- Alyn Corporation

material (company designation): Boralyn NN15

Material 1100 / B4C / 15p / annealed

matrix Al alloy

reinforcement B4C *volume percent* 15

morphology ☒ p ☐ sf ☐ cf ☐ mf ☐ w *size (μm)*

Processing

primary Powder Metallurgy

specifics

secondary ☒ Casting ☒ Forging ☐ Surface Treatments ☒ Extrusion ☒ Machining ☐ Thermal Treatments

specifics

joining ☐ Brazing ☐ Mechanical ☐ Welding ☐ Other

recyclable ☐ Yes ☐ No

Properties are quoted from manufacturer as general indicators only.

Properties *isotropic* ☒ Yes ☐ No

density ($g\ cm^{-3}$)		*tensile*	
CTE ($x10^6\ K^{-1}$)		*yield strength (MPa)*	90
thermal cond. ($Wm^{-1}K^{-1}$)		*ultimate strength (MPa)*	115
electrical cond. (%IACS)		*modulus (GPa)*	
		elongation (%)	22.3

Property Advantages ☐ Density ☐ Strength ☐ Wear ☐ Fatigue ☐ Thermal Expansion ☐ Weldability ☐ Stiffness ☐ Thermal Conductivity ☐ Other

Applications

Products

Sources Froes, F.H. (1999).

z- Bebop

8570 Hamilton Avenue

Huntington Beach CA 92646 USA

tel ++	1 (714) 374 0200
fax ++	1 (714) 374 0268
email	mrbebop@bebop.com
web address	www.bebop.com

This company is no longer active in the MMC industry, as indicated by the "z-" preceding its name.

Bebop had interest in the specific strength of MMC for bicycling components, developing a derailleur pulley and a seatpost. Poor fatigue properties led to discontinuation of the derailleur pulley application and the seatpost never reached the consumer. Bebop did note the interesting anodising characteristics of particle-reinforced aluminium.

z- Cercast

3905 Industrial Boulevard

Montreal, North Quebec H1H 2Z2 Canada

tel ++	1 (514) 322 2371
fax ++	1 (514) 322 1340
email	cercastinfo@castit.com
web address	www.castit.com

This company is no longer active in the MMC industry, as indicated by the "z-" preceding its name.

Cercast developed two particle-reinforced aluminium materials, Stablite™ reinforced with high concentrations of silicon carbide and Litevar™ reinforced with graphite.

z- Cercast

material (company designation): Stablite™

Material A357 / SiC / 65p

matrix Al - 7.0Si - 0.5Mg

reinforcement SiC *volume percent* 65

morphology ☒ p ☐ sf ☐ cf ☐ mf ☐ w *size (μm)*

Processing

primary Infiltration

specifics

secondary ☐ Casting ☐ Forging ☐ Surface Treatments ☐ Extrusion ☐ Machining ☐ Thermal Treatments

specifics

joining ☐ Brazing ☐ Mechanical ☐ Welding ☐ Other

recyclable ☐ Yes ☐ No

Properties are quoted from manufacturer as general indicators only.

Properties *isotropic* ☐ Yes ☐ No

density ($g\ cm^{-3}$)			*tensile*
		yield strength (MPa)	250
CTE ($x10^{6}\ K^{-1}$)	8.46	*ultimate strength (MPa)*	
thermal cond. ($Wm^{-1}K^{-1}$)		*modulus (GPa)*	160
electrical cond. (%IACS)		*elongation (%)*	

Property Advantages ☐ Density ☐ Strength ☐ Wear ☐ Fatigue ☐ Thermal Expansion ☐ Weldability ☐ Stiffness ☒ Thermal Conductivity ☐ Other

Applications microelectronic packaging

Products

Sources

z- Cercast

material (company designation): Litevar™

Material A357 / Gr / p

matrix Al - 7.0Si - 0.5Mg

reinforcement graphite *volume percent*

morphology ☒ p ☐ sf ☐ cf ☐ mf ☐ w *size (μm)*

Processing

primary Infiltration

specifics

secondary ☐ Casting ☐ Forging ☐ Surface Treatments ☐ Extrusion ☐ Machining ☐ Thermal Treatments

specifics

joining ☐ Brazing ☐ Mechanical ☐ Welding ☐ Other

recyclable ☐ Yes ☐ No

Properties are quoted from manufacturer as general indicators only.

Properties *isotropic* ☐ Yes ☐ No

			tensile
density ($g\ cm^{-3}$)		*yield strength (MPa)*	83
CTE ($x10^6\ K^{-1}$)	6.48	*ultimate strength (MPa)*	
thermal cond. ($Wm^{-1}K^{-1}$)		*modulus (GPa)*	
electrical cond. (%IACS)		*elongation (%)*	20

Property Advantages ☒ Density ☐ Fatigue ☐ Stiffness ☐ Strength ☒ Thermal Expansion ☐ Thermal Conductivity ☐ Wear ☐ Weldability ☐ Other

Applications microelectronic PWBs and packages

Products

Sources

z- Disc Brake Australia Pty Ltd

Unit 33, Slough Business Park
Holker Street, Silverwater
Sydney NSW 2128 Australia

tel ++ 61 (2) 9748 0211
fax ++ 61 (2) 9748 0676
email sales@dba.com.au
web address www.dba.com.au

This company is no longer active in the MMC industry, as indicated by the "z-" preceding its name.

Disc Brake Australia developed brake discs from the Ultalite family of MMC materials produced by Cyco. The visual appearance of the brake discs was inferior, and the brake disc exhibited poor longevity when used with a range of brake pad materials. This development ended as being not sustainable in Australia.

z- Ikon/Lexus Cycling Team
1035 Lake Ride Drive

San Marcos CA 92069 USA

tel ++ 1 (760) 598 1138
fax ++ 1 (760) 598 4812
email verbeast@AOL.com
web address www.ikonlexuscycling.org

This organization no longer uses MMC, as indicated by the "z-" preceding its name.

The Ikon/Lexus Cycling Team raced with bicycle frames produced from Boralyn tube sets. Boralyn is the trade name for a family of boron carbide-reinforced aluminium materials produced by Alyn, now bankrupt.

z- Lanxide Corporation

High Performance Materials Group (HPMG), MSE, Inc.
1300 Marrows Road
Newark DE 19711 USA

tel ++ 1 (302) 456 6200
fax ++ 1 (302) 454 1712
email
web address www.lanxide.com

This company no longer exists, thus it is not active in the MMC industry, as indicated by the "z-" preceding its name.

Lanxide Corporation developed a number of technologies for the production of MMC and ceramic matrix composites. In January of 1999, Lanxide ceased to exist, filing for bankruptcy. Advanced Materials Lanxide (AML) was subsequently formed and purchased the intellectual property of Lanxide. MSE, Inc. and Metek AG operated AML as a joint venture, and became High Performance Materials Group (HPMG), now owned solely by MSE, Inc. A number of companies still exercise licencing agreements from technologies developed at Lanxide, including HPMG, Honeywell Advanced Composites, Brembo, and M-Cubed Technologies among others. It is not always clear which companies use Lanxide technologies for production of MMC and which for ceramic matrix composites.

Other portions of the Lanxide family still exist in various forms. Lanxide Electronics Components (LEC), in particular, is now owned by dmc2 Electronic Components (DEC) and remains active in the MMC industry, serving the electronics sector. Lanxide Armor Products, it seems, no longer exists.

z- Lanxide Electronic Components, Inc (LEC)

dmc2 Electronic Components Corporation (DEC)
PO Box 9409, 1300 Marrows Road
Newark DE 19714-9409 USA

tel ++	1 (302) 456 6300
fax ++	1 (302) 456 6223
email	lec@lec-inc.com
web address	www.lec-inc.com

This company no longer exists, as indicated by the "z-" preceding its name.

dmc2 Electronic Components Corporation (DEC) purchased Lanxide Electronic Components in March 2000 and continues to market "Lanxide" materials for the electronics industry in applications ranging from power conversion to telecommunications and aerospace.

z- Lotus Cars Ltd

Potash Lane
Norwich
Norfolk NR14 8EZ UK

tel ++ +44(0)1953 608000
fax ++ +44(0)1953 608300
email group@lotuscars.co.uk
web address www.lotuscars.co.uk

This company is no longer active in the MMC industry, as indicated by the "z-" preceding its name.

In 1995, the Lotus Elise was standard with MMC brake discs from the so-called Lanxide material from Lanxide Corporation. Lotus stopped using Lanxide material for their brake discs due to the bankruptcy of Lanxide Corporation, reverting to conventional cast iron discs. Lanxide trademarks are now controlled by High Performance Materials Group.

z- MacGregor Golf		
1000 Pecan Grove Drive		
Albany GA	31701	USA
tel ++	1 (800) 841 4358	
fax ++		
email	snaru@macgregorgolf.com	
web address	www.macgregorgolf.com	

This company is no longer active in the MMC industry, as indicated by the "z-" preceding its name.

MacGregor marketed putters produced from Boralyn material. Boralyn is the trade name for a family of boron carbide-reinforced aluminium materials produced by Alyn, now bankrupt.

z- Magnesium Elektron

PO Box 23
Swinton
Manchester M27 8DD UK

tel ++ 44 (161) 91 11 000
fax ++ 44 (161) 91 11 010
email elektronsales@british-aluminium.ltd.uk
web address www.magnesium-elektron.com

This company is no longer active in the MMC industry, as indicated by the "z-" preceding its name.

Magnesium Elektron produced Melram, a family of silicon carbide-reinforced magnesium materials that it offered as three products, for sand cast, die-cast and wrought applications. Due to limited demand for such materials, Magnesium Elektron has ceased production of Melram.

z- Magnesium Elektron

material (company designation): Melram

Material ZC61 / SiC / 12p

matrix Mg - 6Zn - 1Cu - 0.5Mn

reinforcement SiC *volume percent* 12

morphology ☒ p ☐ sf ☐ cf ☐ mf ☐ w *size (μm)* 10

Processing

primary

specifics

secondary ☐ Casting ☐ Forging ☐ Surface Treatments ☐ Extrusion ☒ Machining ☒ Thermal Treatments

specifics

joining ☐ Brazing ☐ Mechanical ☐ Welding ☐ Other

recyclable ☐ Yes ☐ No

Properties are quoted from manufacturer as general indicators only.

Properties *isotropic* ☐ Yes ☐ No

			tensile
density (g cm^{-3})	2.00	*yield strength (MPa)*	370
CTE (x10^{6} K^{-1})	18.5	*ultimate strength (MPa)*	398
thermal cond. (Wm^{-1}K^{-1})		*modulus (GPa)*	63
electrical cond. (%IACS)		*elongation (%)*	0.8

Property Advantages ☒ Density ☒ Strength ☐ Wear ☐ Fatigue ☐ Thermal Expansion ☐ Weldability ☐ Stiffness ☐ Thermal Conductivity ☐ Other

Applications

Products

Sources Magnesium Elektron (1993).

z- Osprey Metals Ltd			
Red Jacket Works			
Millands, Neath		SA 11 NJ	UK
tel ++	44 (1639) 634 121		
fax ++	44 (1639) 630 100		
email	spray.form@ospreymetals.co.uk		
web address	www.ospreymetals.co.uk		

This company is no longer active in the MMC industry, as indicated by the "z-" preceding its name.

z- Reebok International Ltd

USA

tel ++
fax ++
email
web address www.reebok.com

This company is no longer active in the MMC industry, as indicated by the "z-" preceding its name.

Reebok was never particularly active in the MMC industry, however, they did market a watch with Boralyn case, promoted for its strength and lightweight. Boralyn is the trade name for a family of boron carbide-reinforced aluminium materials produced by Alyn, now bankrupt.

z- Reynolds Metals Company

555 Guthridge Court

Norcross GA 30092 USA

tel ++	1 (770) 840 6456
fax ++	1 (770) 734 1560
email	carol_lipphardt@kawneer.com
web address	www.rmc.com

This company is no longer active in the MMC industry, as indicated by the "z-" preceding its name.

Reynolds Metals developed a material called Metalmic® (6061/B4C/ p), produced by powder metallurgy. Metalmic® was meant for the nuclear industry in applications that could benefit from the neutron absorption characteristics of boron carbide. Reynolds Metals is owned by Alcoa.

z- Richardson Metals, Inc

1080 Ford Street

Coloradro Springs CO 80915 USA

tel ++	1 (719) 597 0801
fax ++	1 (719) 570 9623
email	experts@richmetals.com
web address	www.richmetals.com

This company is no longer active in the MMC industry, as indicated by the "z-" preceding its name.

Richardson Metals provides extrusion services for the aluminium industry. At the time of writing, Richardson Metals has stopped extruding MMC and concentrates on extrusion of 6000 series aluminium. Richardson Metals is MIL-I-45208A qualified.

z- Specialized Bicycles

15130 Concord Circle

Morgan Hill CA 95037 USA

tel ++ 1 (408) 779 1631
fax ++ 1 (408) 779 6229
email
web address www.specialized.com

This company is no longer active in the MMC industry, as indicated by the "z-" preceding its name.

Specialized sold the Stumpjumper bicycle frame with at least one tube from M2, the Specialized designation of Alcan's W6A.10A (6061/Al2O3/10p). The M2 Stumpjumper is no longer in production.

z- Textron Systems

201 Lowell Street

Wilmington MA 01887 USA

tel ++ 1 (978) 657 2961
fax ++ 1 (978) 657 2930
email asevigny@systems.textron.com
web address www.systems.textron.com

This company is no longer active in the MMC industry, as indicated by the "z-" preceding its name.

Textron was involved in the development and manufacture of MMC for over fifteen years (formerly under the name AVCO Corp.). Initially, Textron explored aluminium and titanium matrix composites, ultimately focussing exclusively on titanium matrices during the last decade of its activity in MMC. During that time, a silicon carbide monofilament-reinforced titanium alloy was qualified for the exhaust nozzle compression link on the General Electric F110 aircraft engine.

In the early part of 2001, Textron decided to withdraw from the MMC business and to focus development efforts on its SCS silicon carbide monofilaments. Textron believed that the level of activity in the MMC market did not justify the capital investment necessary to remain in the MMC business. Textron believes that the monofilament business remains sufficiently strong to justify continued investment, due to the diversity of a market that reaches beyond the MMC industry.

Textron now supplies SCS-6™ (SiC) monofilaments to, among others, Atlantic Research Corporation, formerly a competitor in the titanium-matrix composite business.

z- Textron Systems

material (company designation): TMC

Material Ti-6242S / SiC / mf

matrix Timetal 21S (Ti-6Al-2Sn-4Zr-2Mo)

reinforcement SCS-6™ SiC *volume percent*

morphology ☐ p ☐ sf ☐ cf ☒ mf ☐ w *size (μm)* 142

Processing

primary

specifics

secondary ☐ Casting ☐ Forging ☐ Surface Treatments ☐ Extrusion ☐ Machining ☐ Thermal Treatments

specifics

joining ☐ Brazing ☐ Mechanical ☐ Welding ☐ Other

recyclable ☐ Yes ☐ No

Properties are quoted from manufacturer as general indicators only.

Properties *isotropic* ☐ Yes ☐ No *tensile*

density (g cm^{-3}) 4.10

CTE (x10^{6} K^{-1})

thermal cond. (Wm^{-1}K^{-1})

electrical cond. (%IACS)

yield strength (MPa) 1600

ultimate strength (MPa)

modulus (GPa) 200

elongation (%)

Property Advantages ☒ Density ☒ Strength ☐ Wear ☐ Fatigue ☐ Thermal Expansion ☐ Weldability ☒ Stiffness ☐ Thermal Conductivity ☒ Other

Applications structural components: landing gear, actuators; high temperature structural applications such as turbine engine components; qualified for GE F110 exhaust nozzle compression link

Products

Sources Textron systems (2000).

z- Unidrive Pty Ltd

45-49 McNaughton Road

Clayton Victoria 3168 Australia

tel ++ 61 (3) 9542 4100
fax ++ 61 (3) 9544 8117
email
web address

This company is no longer active in the MMC industry, as indicated by the "z-" preceding its name.

Unidrive ceased producing MMC for propshafts due to the cost of production; the MMC material was replaced with aluminium 6061 for application in the Corvette and Maserati.

Chapter 6

DISCUSSION AND CONCLUSION

6.1 General Remarks

This survey covers 117 companies active in the metal matrix composite industry at the end of 2001 (20 more previously active), producing some 157 materials and 73 products. One can thus conclude without ambiguity that metal matrix composites have entered the engineering world, even within the rather narrow scope of the present survey (given in Section 2.1: these numbers would for example be significantly higher if cemented carbides and diamond tool materials were included). Metal matrix composites have thus reached a first level of industrial maturity.

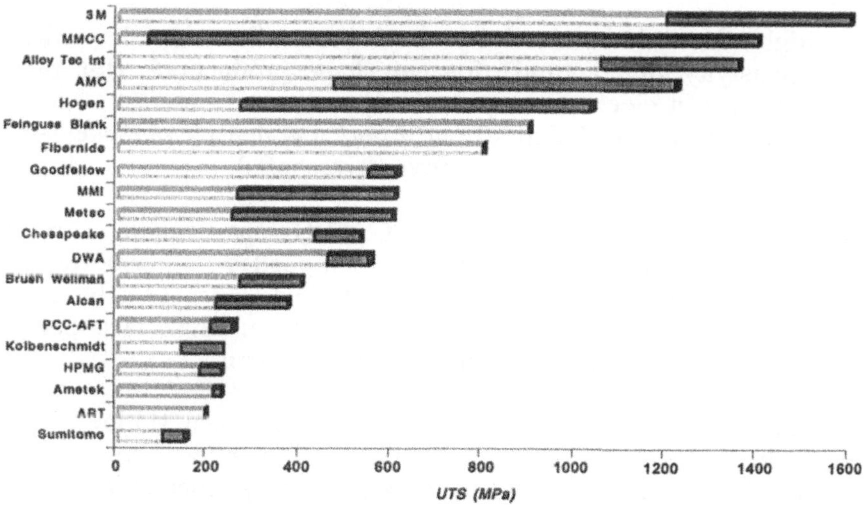

Figure 6.1 - Range of ultimate tensile strengths reported by composite-producing companies surveyed.

As a materials class, industrial metal matrix composite are very diverse: this is clearly apparent in the wide range of properties that can be found in the materials database. An example is given in Figure 6.1, which plots the ultimate tensile strength of commercial metal matrix composites by company: the spectrum of values is very wide indeed (this, and other figures of this chapter, were plotted on the basis of data in the database as it stood shortly before the end of its construction, in November 2001). Similar charts, with an equal spread in values, could be produced for essentially all other properties sampled here: Young's modulus, coefficient of thermal expansion, density, *etc.*, to yield the same conclusion, namely that there is a wide diversity in the properties of the materials surveyed here.

At present, metal matrix composites are used predominantly as "niche" materials in a wide range of diverse applications, ranging from brake discs in trains to critical rotating parts in helicopters, from thermal substrates for advanced electronic and power modules to spikes for track shoes, and from tool materials to combustion engine components. What often drives the use of these materials is not the attraction exerted by one or two general material characteristics alone, but rather the fact that these materials can feature a combination of several properties or attributes that makes them specifically attractive in some applications. Examples of such combinations are:

(i) "high thermal conductivity *and* brazability *and* low thermal expansivity *and* aptitude for net shape production" (for thermal substrates in electronic applications), or
(ii) "high stiffness *and* high strength *and* low density *and* weldability" (for bicycle frames or automotive driveshafts).

Although dominated by SiC and Al_2O_3 particulate reinforced aluminium, commercial metal matrix composites incorporate many different matrices and reinforcement materials. This can be seen in Figures 6.2 and 6.3, which show the range of materials presently combined in metal matrix composites as defined here.

A significant level of specialisation has also been achieved in the MMC industry: producers use a variety of methods to make the composite, Figure 6.4. Also in secondary processing, companies provide a variety of services such as extrusion, secondary casting, and machining, Figure 6.5. This wide range in processes used in metal matrix composite production and secondary processing reflects a dual diversity, in the nature of materials produced and in their end-users.

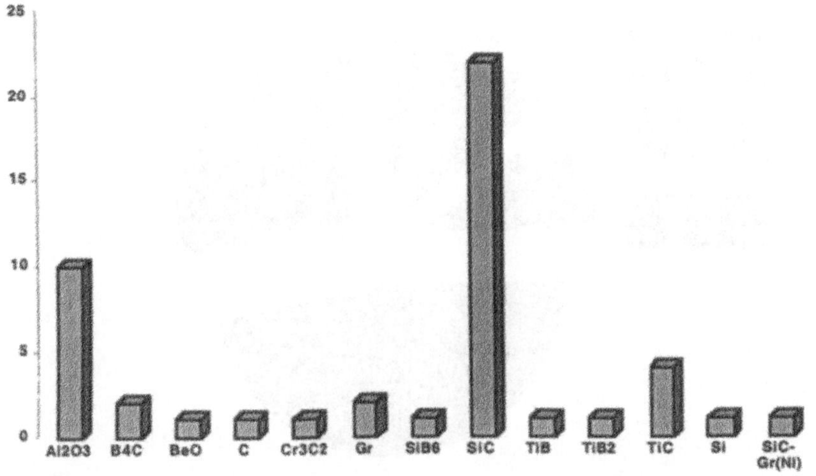

Figure 6.2 - Number of companies in the database according to the reinforcement used in the composites produced.

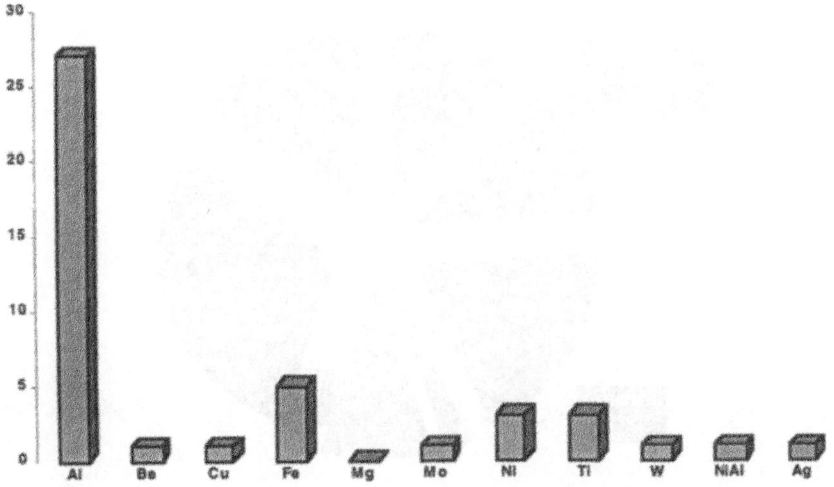

Figure 6.3 - Number of companies in the database according to the base metal of the matrix used in the composites produced.

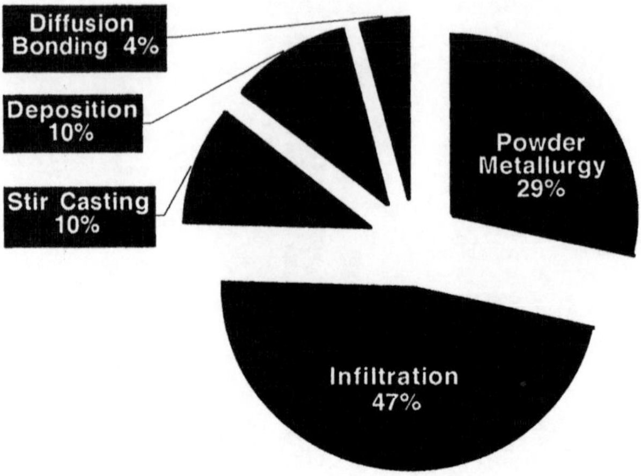

Figure 6.4 - Percentage of companies manufacturing MMC materials by the categorised primary processing methods.

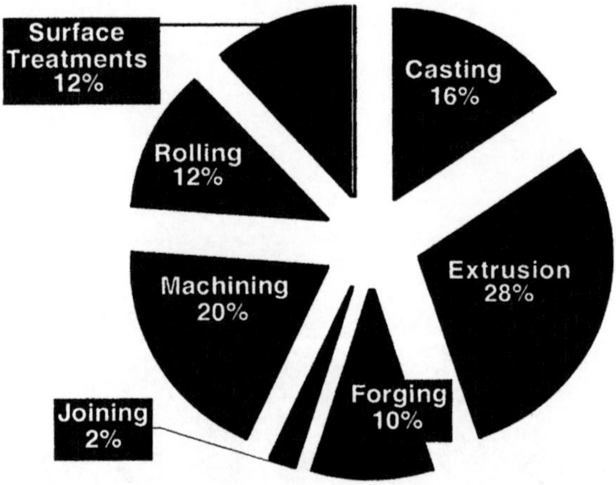

Figure 6.5 - Percentage of companies involved in secondary processing of MMC materials by the categorised processing methods.

Despite this diversity, three sectors predominate in the MMC industry:

(i) automotive and ground transportation,
(ii) aerospace and defence, and
(iii) thermal management for electronics.

Other industries, such as manufacturing and recreational equipment, have also sought the benefits of metal matrix composite materials; however, these are somewhat less prominent.

For applications in ground transportation, price is of primary concern and the benefit of economies of scale is generally sought, particularly in the automotive industry. Companies serving this sector therefore typically use composites produced via liquid metal processes, as these tend to be most cost-effective. In contrast, cost is less of a dominant issue in the aerospace and defence sectors, thus more expensive processes, such as solid state and deposition processes, are often utilized to optimise performance gains. Metal matrix composites for thermal management are widely employed in both ground transportation and aerospace sectors, but are nearly always produced by the liquid metal route, specifically by infiltration, for technical reasons given their high ceramic loadings.

Like most high-tech industries, the MMC industry is dynamic and somewhat volatile. This was apparent in its rapid rate of evolution while this survey was conducted. Over the time-period between December 23 1999 to November 14, 2001, a little over thirty companies that were in the database had left the industry, either by technical choice (*e.g.,* when another material was substituted in a given application for a metal matrix composite) or because of financial constraints including bankruptcy. These companies appear in the database and in Chapter 5 with their name preceded with the letter "z-". About seventy companies were on the other hand added to the database toward the end of that same time period. We have not had time to determine in detail how many of these were new or were simply not found by us at the beginning of the exercise. Hence it cannot be said on the basis of this simple count of the number of companies that have left and entered the present "landscape" to what extent (or even whether) the industry is growing. Nonetheless, this observation certainly shows that this is still a young and "high-tech" branch of industrial activity with a healthy dose of competition from within.

6.2 Automotive and Ground Transportation

Large volume applications of metal matrix composites have been developed largely to serve the automotive and ground transportation industries, exploiting the specific stiffness, high temperature strength, and excellent wear properties of ceramic reinforced aluminium [1]. These applications include selectively reinforced pistons, selectively-reinforced cylinder bores, connecting rods, pushrods, driveshafts, propshafts, brake discs, brake drums, brake callipers, and power module components.

One of the greatest efforts to implement MMC technology in the automotive sector has been in engine components. Selectively-reinforced pistons and cylinder components for diesel engines have been around for nearly two decades and this technology has recently begun filtering into sport automobiles [1-3]. Three Japanese car manufacturers, Honda, Toyota and Mazda, have all developed the manufacturing ability to produce ceramic reinforced metallic engine components. In addition to the Japanese companies, other automotive manufacturers have explored and implemented the technology; *e.g.,* Kolbenschmidt and Porsche with LOKASIL™ infiltrated silicon particle reinforced aluminium cylinder liners [4]. The range and number of MMC components now implemented in explosion engines is far in excess of any other MMC components that have made inroads into the automotive industry. In this class of applications, one could even argue that metal matrix composites are no longer niche materials, but rather belong to the established palette of available engineering materials.

Second to this effort to push MMC technology to the automotive engine sector is the effort made to use MMC materials for automotive brake components. This application is presently growing and MMC brake components are now available on limited production models from the major automobile manufacturers. Brake components, however, are presently restricted to small cars: the Lupo (brake drums on the low-fuel 3L version, available in Europe) from Volkswagen is the highest production automobile containing MMC brake components at present. In addition to added cost of MMC components compared to conventional cast iron, this limitation is related to the physical characteristics of the constituents in the SiC particle-reinforced aluminium typically employed in this application. The relatively low-melting aluminium matrix is susceptible to localised melting if the braking loads are too great, hence presently limiting applications to lightweight cars or to use in the rear wheels (for example as brake drums) where the braking loads are less. In addition, the SiC particulate is highly wear-resistant but also very abrasive, requiring the development of new friction materials for brake pads. On the other hand, aluminium has a higher thermal conductivity than conventional cast iron so that the aluminium-based brakes run comparatively cooler - in addition to being significantly lighter. Ceramic coatings are being developed to be compatible with conventional friction materials and to act as a thermal barrier limiting melting of the aluminium matrix.

Other braking applications include brake fins for the Walt Disney World Big Thunder Railway Thrill Ride, Twin Fin™ composite brakes for use in stationary applications such as the converting industry, and brake discs for motorcycles, bicycles and rail. Knorr-Bremse, for example, has reached the advanced stages of validating MMC brake components for the German high-speed train. Also, the Copenhagen S-Bane (city commuter train) is the first train in the world to have aluminium-based MMC brake discs as standard equipment.

Probably the largest application of metal matrix composites in terms of mass is by General Motors, in its driveshaft application for the extended S/T truck series (the annual production is approximately 150,000 units) [1]. The driveshaft connects and transfers rotational energy from the transmission to the wheels. The extended versions of the S/T truck had two steel driveshafts that were connected with a universal joint; this design was used because of instabilities that would have occurred with a single piece driveshaft. These instabilities are a function of the specific stiffness of the material. A composite having a 6061 aluminium matrix with approximately 20% aluminum oxide provides a thirty-five percent increase in the specific stiffness of the material over aluminium or steel; this is sufficient to allow for a one-piece driveshaft design. GM has also used MMC technology for the propshaft on the Corvette.

Several concept cars, small volume models, racing automobiles and independent carmakers also utilize MMC in these and other applications. WilliamsF1 racing cars reportedly use metal matrix composites in engine components to generate substantial improvements in horsepower [5]. GM's EV-1 is reported to have twelve MMC components, including the engine cradle and the power electronics module; several of these components were supplied by Alyn before it filed for bankruptcy [6]. Tyre studs for Nordic conditions are almost exclusively made from aluminium matrix components since legislation outlaws steel tyre studs [2].

Continuous Al_2O_3 fibre-reinforced aluminium pushrods and brake callipers from 3M are successfully marketed for automotive racing. Due to the competitive nature of the automotive industry, however, information is sometimes difficult to obtain. Relatively small volume applications can be important, nonetheless, as they serve to demonstrate the benefits of MMC components. For example, the first production model automobile with MMC braking technology was the Lotus Elise with MMC brake discs in 1995. Unfortunately this particular application was short-lived due to bankruptcy of the materials supplier; however, other applications of brake discs in automobiles have since followed.

Alcan is the dominant supplier to the ground transportation industries, supplying both cast and extruded forms of its Duralcan material. Hitchiner, Cyco and M^3 Technologies are a few of the names associated with MMC production for the automobile industry, as well as the automakers themselves (Ford, General Motors, Toyota and Honda). Other companies serving the automotive industry include various piston manufacturers, extrusion houses and machine shops. Aluminium industry giants other than Alcan have not succeeded in establishing their metal matrix composite materials in the automotive sector, and some have since ceased developing these materials. Alcoa, for example, discontinued development of particle reinforced aluminium produced by powder metallurgy, and sold their pressure infiltration technology to Polese. Comalco developed the Comral™ MMC, aluminium reinforced with Micral™ spherical alumina-based particles, but discontinued commercialisation some time ago because of slow market development and incompatibility with their primary business operations. Another metal and alloy producer, Norsk Hydro, had also been involved in programs to develop particulate reinforced aluminium for the automotive industry including particle-reinforced aluminium brake discs in conjunction with Volvo Research Corporation [7]. Several of these companies nonetheless remain active in metal matrix composite technology as secondary processers, as shown by the survey in Chapter 5.

6.3 Aerospace Applications

For companies in the defence-related and aerospace industries, property improvements are often the primary concern, such that profits can be achieved even in low-volume production. DWA Aluminium Composites and the companies producing monofilament reinforced titanium composites are particularly active in this sector in applications where specific stiffness is important. Many of these applications are probably carefully guarded secrets; however, some are publicised through large research programs such as Title III in the US [8]. Several important aerospace applications that exploit specific stiffness of MMC have thus been described both in the military and civilian sectors, including ventral fins and fuel access covers in F-16 aircraft, the blade sleeve and swash plate on helicopters from Eurocopter, floor support struts in the Airbus, and fan exit guide vanes on Pratt-Whitney engines for the Boeing 777 aircraft (the use of MMC in this last application is quoted to produce savings of $ 100 million over 10 years, and double this figure in life cycle savings) [2,8].

The aerospace industry faces the challenge of qualifying new materials before they can be implemented. This qualification process relies, in part, on developing a detailed understanding of failure mechanisms associated with the material; this has been commonly cited as a hindrance to widespread application of MMC in the aerospace industry. Nevertheless, extensive work continues on a diverse range of MMC materials to serve the military and aerospace sectors, both in the US and in Europe. The view that MMC can be qualified is illustrated by inclusion of MMC in the U.S. Military Handbooks, the material design guide for the aerospace industry: MIL-HDBK-17-4 is devoted to MMC. DWA has qualified particle-reinforced aluminium anticipating inclusion in MIL-HDBK-5 (Metallic Materials and Elements for Aerospace Vehicle Structures). Rolls-Royce in the UK has reportedly also conducted extensive qualification of their own particle-reinforced aluminium. Continuous fibre-reinforced titanium, in particular, has been extensively tested as demonstrated by the volume of data published in MIL-HDBK-17-4.

The space industry has a history of utilising metal matrix composites because it pushes the boundaries of engineering materials with less concern for component cost [9,10]. Demands in this sector often focus on the dimensional stability of materials under the inevitable thermal cycles associated with passing in and out of the Sun's shadow in the absence of convective heat transfer, in addition to absolute weight savings to reduce launch costs. Therefore low thermal expansion coefficient, low density, high stiffness, and high strength are desired. Metal matrix composites have been utilized on several high-profile space projects. The Space Shuttle Orbiter utilises continuous boron fibre reinforced aluminium tubular struts as frame and rib-truss members of the mid-fuselage section, and as the landing gear drag link; this is the first successful MMC application in space. The high-gain antenna boom for the Hubble telescope is another application of MMC in space; the boom is made from carbon fibre reinforced aluminium composite produced by diffusion bonding. The material satisfies the principal requirements of high specific stiffness and low (near zero) CTE, as well as transmitting electrical signals to the antenna dish due to its excellent electrical conductivity.

Metal matrix composites are also used for electronic packaging and thermal management applications in spacecrafts and satellites, for example in Motorola's Iridium satellite network. Beryllium-based composites are another material used in space, notably for satellites, due to the exceptional weight savings obtained by using these materials, despite the very high cost of the raw material and the health hasards it may pose.

6.4 Thermal Management in Electronics

Metal matrix composites used by the electronics industry have a significantly different character than the MMC used for structural applications because the electronics industry exploits mainly the dilatational and thermal properties of the MMC, with less interest in its mechanical properties [11,12]. Thus these MMC materials are often more ceramic than metal by volume: the ceramic particle volume fraction is generally around 70%, although materials with lower fractions of ceramic are suitable for some applications. AEA, Ametek, Ceramic Process Systems (CPS), Electrovac, Enertron, dmc2 Electronic Components (formerly Lanxide Electronic Components), MC-21, Olin Aegis, PCC-AFT, Polese, Siemens, Sumitomo Electric and TWM Technology are some of the companies reportedly active with Al-SiC composites for electronic applications. These applications include thermal cores, chip carriers, heat sinks and thermal spreaders, microprocessor lids and microwave packages; therefore, MMC for thermal management can be found on passenger trains, automobiles, aircraft, and spacecraft. This application is quoted to represent over one million parts annually [10].

6.5 Other Applications

Most of the applications mentioned above make use of discontinuously reinforced metals, mainly ceramic particle-reinforced aluminium. The principal continuous fibre reinforced metal (CFRM) with an aluminium matrix commercially available today is from 3M which produces pushrods for auto racing, in addition to reinforced wire, stiffeners and rods (Feinguss-Blank and MMCC apparently have produced CFRM, but it seems that only 3M actually sell the material). 3M, in collaboration with EDF in France, is testing continuous alumina-fibre reinforced aluminium for use in overhead power transmission cables. If successful and qualified, this application has the potential of dramatically increasing the level of production of this material, in turn lowering the price and paving the road for additional applications. A few companies are involved in monofilament reinforced composites with titanium matrices although it is not known whether actual applications exist outside the defence industries.

Although smaller in number, there are activities related to steel matrix composites. Alloy Technology International, Alloytic and Edelstahl

Witten Krefeld market a variety of iron- and nickel-based alloys reinforced with 30 to 45 volume percent TiC for a diverse range of applications, particularly for the manufacturing industries. Metco Powdermet Oy make steel reinforced material that has applications including cladding for erosion resistance in riser tubes for the petrochemical industry. AMC is also developing particle-reinforced steel composites.

The sports industries have always embraced so-called aerospace materials, in part because the competitive nature of sport assures that the consumer is willing to pay a premium for equipment utilizing high-performance materials. In this industry, even rumour can influence materials choices and clever marketing is often more important than engineering. In addition to bicycle frames, several bicycle components that are manufactured from particle-reinforced aluminium have been marketed at one time or another: derailleur pulleys, seatposts, and rims. Golf clubs and putters are available from several companies, and one company reportedly produces lacrosse sticks. Reebok even has offered a watch with a MMC casing. MMC shoe spikes from Omni-Lite Industries were introduced in the Olympics at Atlanta, claiming 20 Gold medals in track and field events. A number of additional products have been developed, but production has not been verified; these include showshoes, baseball bats, and ski poles amongst many others.

6.6 Conclusion

It has been mentioned in recent years, in a humouristic sense of course, that "metal matrix composites are no longer materials of the future". In a way, this is true: metal matrix composites have indeed entered the engineering world. They represent a wide array of materials, which are commercially available in a range of forms and from several (often competing) manufacturers. Their applications are variegated and include mass-produced components. Their use has also in several cases been documented to represented significant savings or to produce impressive performance improvements. In short, metal matrix composites have become "real" industrial materials.

Still, these composites remain "cutting-edge". Although not a new concept in a few specific materials systems (*e.g.*, copper/tungsten or cemented carbide), the idea of producing a composite material having a metal and not a polymer as its matrix is generally considered unusual, as is the idea of "artificially" reinforcing a metal with another material such as a ceramic fiber.

There remains also much to be proven and explored with these materials. This is apparent for example if one measures the tremendous progress that has been achieved during the past ten years in some areas of this technology. Examples include ceramic fibres, composites for thermal management, or development work in some specific MMC application systems such as brake rotors or electronic substrates. The perception that these materials are "cutting-edge" or "high-tech" is thus accurate.

Will these materials remain niche materials ? To some extent, and although this view is not shared by all who are familiar with metal matrix composites (*e.g.*, [8]); we believe that metal matrix composites are niche materials, and will remain so at least during the next decade. One reason is that, by essence, metal matrix composites are "designer" materials that can be tailored to a significant degree to meet the specific needs of the end-user. These materials are also more expensive than metals and alloys; hence their implementation in engineering applications will generally require that they provide "something extra" in terms of properties and performance, which will be specific for each application. As a consequence of these two features, their inroads into engineering usage will nearly every time happen for very specific reasons: these traits essentially define a "niche material".

But several of these niches are already quite large and significant. Components at the heart of explosion engines, such as pistons or cylinder materials, are quite a big "niche" indeed, as are electronic packaging materials. These niches already add up to quite a respectable array of engineering applications, and are growing in size and number. Furthermore, with each new application, engineering experience is gained, which can significantly reduce the barrier that must be overcome for the next application, even if it is each time rather specific.

Whether metal matrix composites will remain "niche" materials is, thus, to us not much of an issue: the potential they hold in tangible applications is nonetheless high. Rather, we believe one of the main barriers to their more general use is the simple need, for designers, to know and be able to access "who is doing what" in engineering with metal matrix composites, and also to gain rapid and simple access to what these materials are and how they are made without having to read the very large volume of research literature on metal matrix composites. We hope this volume, which is aimed at filling this gap, can contribute somewhat towards making this class of materials just a bit more banal.

APPENDIX A - Glossary of terminology commonly used in relation to metal matrix composites

Aspect ratio: ratio of length to diameter of short fibres or whiskers.

Binder: an additional phase added to the reinforcement to create a preform, with the goal of increasing the tensile and shear strength of the preform for handling prior to infiltration. Colloidal silica is the most common binder for preforms, typically applied wet prior to shaping the preform, which is subsequently dried by firing. Binders may also be used in powder metallurgy methods to increase handling strength or to lubricate flow (in injection molding for example); they are then burned off prior to consolidation.

Borsic: silicon carbide coated boron monofilaments.

Cellular metals (metal foams): highly porous metallic material; may be considered as MMC formed of a metal matrix "reinforced" by closed or open pores containing vacuum or gas (these are not considered composites in the present document).

Centrifugal infiltration: an infiltration process in which centrifugal force is used to drive the molten matrix into the preform; typically, high angular velocities (several thousand rpm) are needed.

Cermet: a metal matrix composite with a three-dimensionally continuous ceramic reinforcement, typically with more ceramic than metal (generally containing less than 20% metal by volume). A cermet is thus both a ceramic matrix composite and a metal matrix composite.

Chopped fibres: short fibres produced by cutting a fibre into shorter discrete lengths.

Coating processes: see Deposition processes

Cold isostatic pressing (CIP): cold pressing of composites with application of hydrostatic pressure by means of a non-compressible fluid such as oil, at temperatures too low to effect concurrent sintering of the matrix; for metal matrices this temperature is generally less than the matrix recrystallisation temperature.

Cold pressing of composites: powder consolidation of composites with application of pressure by means of a piston, at temperatures too low to effect concurrent sintering of the matrix, which for metal matrices is generally below the matrix recrystallisation temperature.

Compocasting: a primary composite manufacturing process, a variant of stir-casting in which the metal is semi-solid, *i.e.*, a rheocasting process incorporating the reinforcing ingredient material to form the composite. This aids incorporation of the particles, in particular because the metal is more viscous.

Composite consolidation: a primary composite manufacturing process in which the ingredient materials comprising matrix and reinforcement elements (generally the reinforcement coated with the matrix metal) are consolidated, with or without pressure, to eliminate porosity.

Composite: a composite (or composite material) is defined as a material that consists of at least two distinct phases (or combinations of phases) that are bonded together at the atomic level in the composite, each of which originates from a separate ingredient material that pre-exists the composite. The essential elements of this definition are: (i) composite refers to a material, as opposed to a structure or a component; as such a composite material is used for the fabrication of components of various shapes or functions, which distinguishes it from a wing or other structure made of several components bonded together, or from an electronic device or packaging structure made of layered materials (although one of the materials in the packaging could be considered a composite), and (ii) the composite is produced via a physical combination of at least two pre-existing ingredient materials; this distinguishes a composite from other multiphase materials which are produced by bulk processes where one or more phases result from phase transformation: directionally solidified eutectics or molten alloys from which a ceramic phase is solidified or precipitated are thus considered to be alloys and not composites (even though their properties can be described using composite theory).

Continuous fibre (cf) : a cylindrical ingredient material produced continuously to form an essentially endless reinforcement in the composite; usually delivered on bobbins of fibre tows, each tow consisting of many individual fibres of diameters typically in the range of 5 to 20 μm. In the production process such fibres are often coated by a polymeric **size** or **sizing.**

Continuous fibre reinforced MMC (CFRM): a metal matrix composite with reinforcement of continuous fibres.

Continuous MMC primary process: a primary process for the production of MMC material from its ingredients in continuous fashion (*e.g.*, foil-fibre-foil diffusion bonding using a rolling mill, or continuous infiltration).

Continuous reinforcement: a reinforcement that continues across the composite in at least one dimension (typically made of continuous fibres or an open-celled foam).

Deposition processes: processes that deposit (or coat) the matrix onto individual reinforcement elements of the ingredient material. Examples include vapour deposition and electrolytic coating.

Diffusion bonding: a primary process for composite fabrication in the solid state, in which stacks of reinforcement preforms (usually parallel fibre monolayers) and matrix foils are stacked, encapsulated, and bonded together after evacuation by mechanical pressure.

Discontinuous reinforcement: a non-continuous reinforcement, taking the form of individual elements embedded in the metal matrix (e.g., particulates, short fibres, whiskers).

Discontinuously reinforced MMC: a metal matrix composite with a discontinuous reinforcement.

Dispersoid (or dispersion) reinforced MMC: a metal matrix composite with a dispersoid reinforcement occupying a volume fraction in the material greater than 5% (otherwise, the material is considered a dispersion strengthened metal; incidentally it may form the matrix of any type of MMC, *i.e.*, a MMC with dispersion-strengthened matrix).

Dispersoids (d): same as particulates except for the diameter, which is less than 1 μm.

Electrolytic co-deposition: a primary composite processing method that consists in electrolytic deposition of the matrix metal from a solution containing the reinforcement ingredient, so as to trap the reinforcement within the deposited metal layer.

Electrolytic coating: a primary composite manufacturing process wherein the matrix is deposited from solution using electrochemical means onto individual reinforcement elements.

Flake: a flat roughly equiaxed reinforcement. Graphite is, for example, often supplied in flake form (see platelet).

Foil-fibre-foil method (also called **foil-fibre method**): process wherein layers of metal foil and ceramic monofilament fibres are alternatively stacked to produce precursor materials and subsequently hot pressed, also called diffusion bonding.

Forced infiltration: an infiltration process which is not spontaneous, *i.e.,* which requires application of mechanical force onto matrix or reinforcement.

Gas pressure infiltration: a pressure infiltration process that uses pressurized gas to apply pressure on the metal surface.

Hot isostatic pressing (HIP): hot pressing of composites with application of hydrostatic pressure by means of a gas, at temperatures high enough to effect concurrent sintering of the matrix, which for metal matrices is generally above the matrix recrystallisation temperature.

Hot pressing: powder consolidation of composites with application of pressure by means of a piston, at temperatures high enough to effect concurrent sintering of the matrix, which for metal matrices is generally above the matrix recrystallisation temperature.

Hybrid composite: a composite consisting of at least three (as opposed to two) distinct phases (or combinations of phases) that are bonded together at the atomic level in the composite, each of which originates from a separate ingredient material that preexists the composite (there were, hence, at least three ingredient materials).

Hybrid preform: a preform containing at least two types of ingredient materials that form at least two distinct reinforcements in the composite.

Infiltration: a primary process of composite production whereby the molten metal matrix is made to fill, spontaneously or under force (*i.e.*, with supply of external mechanical work), pores within a preform of the reinforcement.

Ingredient material supplier or producer: a supplier or producer of ingredient materials that are used to make a composite (includes fibre or ceramic particle producers, as well as preform producers).

Ingredient materials: raw materials put together to produce the composite (*e.g.*, the metal alloy, metal powder, ceramic particles, etc.)

In-situ MMC: a MMC in which the ingredient reinforcement material changes its nature during composite processing (*e.g.*, TiB_2 reacts to form TiC in some metals by adding carbon). To obtain a MMC, the reinforcement must, however, retain its identity and in particular remain solid throughout processing of the reinforcement (otherwise, the material is a solidified alloy).

Interface: the portion of the composite microstructure that lies between matrix and reinforcement. The interface may be a simple row of atomic bonds (*e.g.*, the interface between alumina and pure aluminium), but may also include matrix/-reinforcement reaction products (*e.g.*, aluminium carbide between aluminium and carbon fibres), or reinforcement coatings (*e.g.*, interfacial coatings between SiC monofilaments and titanium matrices).

Interpenetrating phase composite (IPC): a composite in which both matrix and reinforcement phases are three-dimensionally continuous (in the sense that a line of a certain finite thickness can traverse the composite in all three directions without leaving either the matrix or the reinforcement) throughout the material. IPCs are produced for example by pressure-infiltration with molten metal of an open-pore ceramic foam or of a sintered ceramic powder preform.

Liquid phase sintering: powder consolidation of composites in the (at least temporary) presence of a liquid phase in the powder compact.

Local reinforcement: see selectively reinforced metal or composite).

Longitudinal property (*such as modulus, strength, thermal expansion,*

etc): the value of a property along the direction of aligned fibres or monofilaments.

Lorentz force infiltration, or **electromagnetic infiltration**: an infiltration process in which electromagnetic forces push the molten matrix into the preform.

Matrix coated fibre method: a type of composite consolidation process whereby continuous fibres or monofilaments are coated with metal matrix material prior to composite consolidation; coating may be by physical vapour depostion, or alternatively by sputtering techniques or by chemical means.

Mechanical pressure infiltration: a pressure infiltration process that uses a moving solid part to apply pressure on the metal surface.

Metal Matrix Composite (MMC): a composite material in which a metal or alloy forms a continuous network.

Metal matrix: the continuous metal or alloy in a metal matrix composite that defines it as a MMC .

MMC component supplier or producer: a supplier or producer of metal matrix composite material components produced to final shape and microstructure (the producer performs at least secondary processing of the material).

MMC consumer: the user of an end-product that incorporates metal matrix composite materials (for example the cyclist with reference to a bicycle).

MMC designation: a method of indicating the composition (and in some cases the processing) of the MMC. Designations vary by company, although there exists a proposal for designation of aluminium composites by the Aluminum Association. We follow in the database the designation used by the companies, and in our own writings, we designate the composite as follows: accepted designation of the matrix / abbreviation of the reinforcement's designation / arrangement and volume fraction in % with symbol of type (shape) of reinforcement (*e.g.*, AA6061 / Al2O3 / 22 p).

MMC end-user: the producer of an end-product that incorporates metal matrix composite materials (for example a bicycle manufacturer)

MMC insertion casting: a secondary process for production of partially reinforced components, in which a pre-fabricated MMC is placed within a mold, and incorporated into a metal or alloy by a casting process.

MMC supplier or producer: a supplier or producer of metal matrix composite materials (producer performs at least primary processing of the material).

Monofilament reinforced (MFRM): metal matrix composite with reinforcement of monofilaments.

Monofilaments (mf): same essentially endless type of reinforcement as continuous fibres, except for a larger diameter, typically more than 100 μm. Monofilaments are generally produced by deposition onto a core fibre (most often of carbon or tungsten), and are delivered as individual fibres instead of tows.

Partially reinforced metal or composite: see selectively reinforced metal or composite).

Particulate reinforced metal (PRM): metal matrix composite with a particulate reinforcement occupying a volume fraction in the material greater than 5% (otherwise, the particulate are generally considered to be inclusions).

Particulates (p): roughly equiaxed reinforcement or composite ingredient, usually of aspect ratio (ratio of largest to other two perpendicular diameters) less than about 5. Particulates can be both mono- or polycristalline, can take various shapes (spherical, angular, plate-like) and are typically greater than 1 μm in diameter.

Plasma spray method: primary processing method, variant of the foil-fibre-foil method where monolayers of monofilament fibres or continuous fibre tows are plasma sprayed with the matrix metal and subsequently stacked and hot-pressed.

Platelets: flat reinforcements, generally of equiaxed cross-section within their plane, and most often single-crystals (their shape being derived from the growth characteristics of the crystal).

Powder cloth method: variant of the foil-fibre-foil method for pre-production of monofilament reinforced metal, in which matrix metal powders and a binder are mixed and rolled or slurry-cast into a cloth or foil before stacking and subsequent hot pressing.

Powder consolidation: a primary process for composite fabrication in which matrix in powder form is blended or otherwise placed next to the reinforcing phase elements and consolidated, with or without pressure, to eliminate porosity.

Powder extrusion: powder consolidation of composites with application of pressure by means of an extrusion press. **Cold** extrusion is below the metal recrystallisation temperature, **hot** extrusion above.

Powder forging: powder consolidation of composites with application of pressure by means of a forging press. **Cold** forging is below the metal recrystallisation temperature, **hot** forging above.

Powder pressing: powder consolidation of composites with application of pressure by means of a piston.

Powder rolling: powder consolidation of composites with application of pressure by means of a rolling mill. **Cold** rolling is below the metal recrystallisation temperature, **hot** rolling above.

Pre-processing: all steps which precede primary processing (*e.g.*, surface treatment of ingredient materials, or preform fabrication for infiltration processing).

Preform: a shaped porous assembly of ingredient material elements (such as fibres, whiskers or particles). Typically, preforms are produced for subsequent infiltration with liquid metal, or in fibre layer / metal foil / fibre layer stacks prior to diffusion bonding. The mechanical stability of the shape may be provided by the adjunction of a **binder**, potentially but not always fugitive, and amounting to only a few mass % of the preform.

Pressure die casting (UK terminology) or **Die casting** (US terminology): a casting process in which a piston pushes the melt into a solid mould causing turbulent flow.

Pressure die infiltration: an infiltration process applying pressure die casting, in which a preform is placed into a solid mould, and pressure infiltration is effected with a moving solid piston.

Pressure infiltration: an infiltration process in which hydrostatic pressure is applied onto the molten matrix surface to drive it into the preform.

Primary processing: production of the composite material by combining its ingredient materials, but not necessarily to final shape or final microstructure.

Producer: an industrial or research institution that produces MMC or MMC ingredient materials, or products thereof.

Random planar: a term used to describe the orientation of short fibers within a preform, and meaning that their axes are random, but restricted to a single plane. Generally, this is the plane perpendicular to the direction along which the fibers were pressed when preparing the preform.

Reinforcement: a phase or combination of phases originating from the ingredient material which is combined with a metal or an alloy to produce a metal matrix composite (*e.g.*, alumina fibres, silicon carbide whiskers, steel fibres, or graphite particles, even if this last reinforcement does not "reinforce" the matrix). A reinforcement is characterised by its chemical composition, its shape and dimensions, its properties as ingredient material and its volume fraction and spatial distribution in the matrix.

Rheocasting: a family of metal alloy or composite processes that involve the solidification of stirred semi-solid metal (typically, the resulting semi-solid slurry exhibits strongly thixotropic behaviour).

Roving: a bundle of fibre tows (twisted or untwisted).

Secondary processing: processing steps that follow primary processing, and aim to alter the shape or microstructure of the material (*e.g.*, shape casting, forging, extrusion, heat-treatment, machining).

Selectively reinforced metal or composite: semi-product or component that contains MMC material metallurgically bonded to a conventional metal or alloy (*e.g.*, by diffusion bonding, insertion casting, or co-extrusion). The result is a material or a component of a given metal or alloy that is only reinforced over a portion of its volume. A typical example is that of Al-Si automotive components produced by pressure casting and infiltration, which are reinforced only in certain areas (such as the ring land areas or cylinder liner surfaces) with ceramic short fibres or particles.

Semi-solid forming: a family of metal alloy or composite forming processes that use semi-solid metal.

Short fibre reinforced metal (SFRM): metal matrix composite with a short fibre reinforcement.

Short fibres (sf): non-continuous reinforcement or composite ingredient with a ratio of length to diameter above 5 and of diameter typically above 1 μm.

Shot: roughly spherical ceramic inclusions found in many short fibre preforms, larger than the fibres in diameter, and which are an unwanted by-product of the fibre manufacturing process. These tend to reduce composite mechanical properties (in particular with regard to fatigue strength, as they serve as crack initiation sites). It is difficult to purchase shot-free ceramic short fibres because these fibres are generally mass-produced as thermal insulation, and only sold in marginal quantities for metal matrix composite reinforcement.

Sintering: powder consolidation of composites without application of pressure.

Sizing, or Size: a thin coating (typically polymer) on many continuous fibres designed to prevent fiber surface damage during winding and handling, and to optimize the interfacial bond (generally, however, sizings found on commercial fibers are designed for the production of polymer, not metal, matrix composites).

Slurry tape casting: a pre-processing method where monofilaments are coated with a slurry of matrix metal powder thus forming a "tape" that can be stacked and hot pressed to form MFRM.

Spontaneous infiltration: an infiltration process in which the matrix penetrates the preform spontaneously without added force, either naturally (*e.g.*, copper into tungsten), or artificially (*e.g.*, by coating nickel onto the reinforcement for infiltration with aluminium, or by control of the atmosphere and the alloy composition as in the Lanxide Primex™ process).

Spray casting: a primary process of composite production whereby the metal is sprayed onto a substrate. For composites, the reinforcement is either already incorporated in the sprayed melt (*e.g.*, by stir-casting), is combined during spraying with the metal, by injection of the reinforcement ingredient material into the sprayed metal droplet stream, or is co-sprayed, *i.e.*, sprayed at the same time as the matrix onto the substrate.

Spray deposition: a primary process similar to spray casting except that the substrate is the reinforcement, *i.e.*, the matrix is sprayed on to the reinforcement typically as monolayers which are subsequently consolidated.

Squeeze casting: a casting process in which a piston pushes molten or semi-solid metallic material into a solid mould. Variants include **direct squeeze casting**, in which the piston surface represents a fraction of the final casting surface; and **indirect squeeze casting**, in which the piston acts on the metal outside the die providing laminar flow of the melt into the die.

Squeeze casting infiltration: a mechanical pressure infiltration process in which a preform is placed into a solid mould, and pressure infiltration is affected with a moving solid piston providing laminar flow of the molten matrix into the mould.

Staple fibres: short fibres

Stir casting: a primary process of composite production whereby the reinforcement ingredient material is incorporated into the molten metal by stirring the latter in contact with the former.

Supplier: an industrial or research institution that markets and sells MMC or MMC ingredient materials, or products thereof.

Thixocasting: a family of metal alloy or composite casting or forming processes which shape partially remelted thixotropic semi-solid metallic materials by deformation (thixotropy refers to materials that have a time dependent shear strength, and many metal alloys and composites display this behaviour in the semi-solid state).

Tow: a bundle of continuous fibres, typically several thousand; they are produced in such bundles and wound on a spool for delivery.

Transverse property (*such as modulus, strength, thermal expansion, etc*): the value of a property along a direction normal to aligned fibres or monofilaments.

Ultrasonic infiltration: an infiltration process in which ultrasonic pressure waves are used to drive the molten matrix into the preform.

Vacuum infiltration: an infiltration process in which a hydrostatic pressure of one bar applied by the atmosphere onto the molten matrix drives it into an evacuated preform. The vacuum can in some processes be **self-generated**, for example by reaction of magnesium with air or oxygen within the preform.

Vapour deposition (in the context of composite production): a primary process for composite production wherein the matrix is deposited from the vapour phase onto individual reinforcement elements of the ingredient material. Vapour deposition onto a thin carbon or tungsten fibre is also used to produce monofilaments.

Vapour infiltration processing: a primary process for composite production wherein the matrix is gradually deposited from the vapour phase into open pores of a reinforcement preform (generally used for ceramic matrix composite production).

Whisker reinforced metal (WRM): metal matrix composite with a whisker reinforcement.

Whiskers (w): elongated single crystals, typically produced with a length to diameter ratio greater than 10 and of diameter usually less than 1 μm.

Wire: a continuous fibre or monofilament made of metal, typically of diameter equal to or greater than 100 μm.

Wire winding: a pre-processing method for production of monofilament reinforced metal which consists in co-winding metal wires and ceramic fibres for subsequent hot pressing.

APPENDIX B - Database keywords

1 - COMPANIES

Composite producers
Manufacturer of components
Supply-house
End-user
Other

Primary processing
Deposition
Diffusion bonding
Infiltration
Liquid phase sintering
Plasma spray
Powder metallurgy
Spray casting
Stir casting
Other

Secondary processing
Casting
Extrusion
Forging
Joining
Machining
Rolling
Thermal treatment
Surface treatment
Other

Other keywords
Country (in address): Australia, Austria, Belgium, Canada, Finland, France, Germany, Israel, Italy, Japan, Sweden, Switzerland, UK, USA.

2 - MATERIAL

Material

Matrix: searchable by main matrix alloy element: Al, Fe, Cu, Ni, Mg, Ti, Be, ...

Reinforcement: searchable by
- the four categories: p = particulate, sf = short fibre, w=whisker, cf =

continuous fibre, mf = monofilament, and
• chemical identity from among Al2O3, TiC, SiC, B, B4C, BeO, Gr (graphite) and SiB6.

Property advantages
Isotropic

Primary processing
Deposition
Diffusion bonding
Infiltration
Liquid phase sintering
Plasma spray
Powder metallurgy
Spray casting
Stir casting
Other

Secondary processing
Casting
Extrusion
Forging
Machining
Surface treatments
Thermal treatments
Other

Industries:
Aerospace
Automotive and Transportation
Chemical, Gas and Oil
Communications
Defence
Electrical and Electronics
Manufacturing
Mining, Iron and Steel
Power
Recreation
Other

Applications:
Braking: wear
Coatings/Linings
Combustion Engines: wear

Neutron Absorption
Structural
Structural: high-temperature
Structural: sheet
Structural: tubing
Therm Management: electronics
Wear/Corrosion Resistance
Other (search text box)

3 - PRODUCTS

"Component" and "Description"
various terms, including
Bicycle
Blade
Brake
Brake
Conductor
Crankshaft pulley
Cylinder
Die
Diesel
Driveshaft
Electronic packaging
Engine cradle
Fan Exit Guide Vanes
Fin
Fuel access covers
Golf
Heat spreader
Helicopter
Lacrosse
Landing gear
Manifold
Piston
Propshaft
Pushrod
Rack
Satellites
Shoe
Stiffeners
Tyre
Valve
Watch

Form
Casting
Cladding
Coating
Drawn
Extrusion
Forging
Sheet

BIBLIOGRAPHY AND REFERENCES

A - Bibliography

The most up-to-date and complete source of information on metal matrix composites is the recent encyclopedia published by Elsevier, titled *Comprehensive Composite Material,* edited by A. Kelly and C. Zweben, and in particular its Volume 3, devoted to metal matrix composites and edited by T.W. Clyne. Among many relevant articles in this six-volume encyclopedia, the following cover directly the materials, their processing and their applications. Many other chapters in this work also provide detailed coverage of microstructure - property relations in metal matrix composites.

Clyne T.W.: "An Introductory Overview of MMC Systems, Types, and Developments, Chapter 3.01" in *Comprehensive Composite Materials, Vol. 3: Metal Matrix Composites*, T.W. Clyne, ed., Pergamon, Oxford, UK, 2000, pp. 1-26.

Prangnell P.B.: "Precipitation Behaviors in MMCs, Chapter 3.03" in *Comprehensive Composite Materials, Vol. 3: Metal Matrix Composites*, T.W. Clyne, ed., Pergamon, Oxford, UK, 2000, pp. 61-90.

Mortensen A.: "Melt Infiltration of Metal Matrix Composite, Chapter 3.20" in *Comprehensive Composite Materials, Vol. 3: Metal Matrix Composites*, T.W. Clyne, ed., Pergamon, Oxford, UK, 2000, pp. 521-554.

Lloyd D. and Jin I.: "Melt Processed Aluminum Matrix Particle Reinforced Composites, Chapter 3.21" in *Comprehensive Composite Materials, Vol. 3: Metal Matrix Composites*, T.W. Clyne, ed., Pergamon, Oxford, UK, 2000, pp. 555-577.

Li B. and Lavernia E.: "Spray Forming of MMCs, Chapter 3.23" in *Comprehensive Composite Materials, Vol. 3: Metal Matrix Composites*, T.W. Clyne, ed., Pergamon, Oxford, UK, 2000, pp. 617-653.

Ward-Close C.M., Robertson J.G. and Godfrey S.P.: "Fabrication of Monofilament Reinforced Titanium, Chapter 3.24" in *Comprehensive Composite Materials, Vol. 3: Metal Matrix Composites*, T.W. Clyne, ed., Pergamon, Oxford, UK, 2000, pp. 679-700.

Gheorghe I. and Rack H.J.: "Powder Processing of Metal Matrix Composites, Chapter 3.25" in *Comprehensive Composite Materials, Vol. 3: Metal Matrix Composites*, T.W. Clyne, ed., Pergamon, Oxford, UK, 2000, pp. 679-700.

Hunt W.H.: "Particulate Reinforced MMCs, Chapter 3.26" in *Comprehensive Composite Materials, Vol. 3: Metal Matrix Composites*, T.W. Clyne, ed., Pergamon, Oxford, UK, 2000, pp. 701-715.

Maruyama B.: "Continuously Reinforced MMCs, Chapter 3.27" in *Comprehensive Composite Materials, Vol. 3: Metal Matrix Composites*, T.W. Clyne, ed., Pergamon, Oxford, UK, 2000, pp. 717-739.

Miracle D.: "Intermetallic Matrix Composites, Chapter 3.28" in *Comprehensive Composite Materials, Vol. 3: Metal Matrix Composites*, T.W. Clyne, ed., Pergamon, Oxford, UK, 2000, pp. 741-778.

Bader M.G.: "The Composites Market, Chapter 6.01" in *Comprehensive Composite Materials, Vol. 6: Metal Matrix Composites* M.G. Bader, K. Kedwards and Y. Saweda, eds, Pergamon, Oxford, UK, 2000, pp. 1-13.

Hunt W.H.: "Metal Matrix Composites, Chapter 6.05" in *Comprehensive Composite Materials, Vol. 6: Metal Matrix Composites* M.G. Bader, K. Kedwards and Y. Saweda, eds, Pergamon, Oxford, UK, 2000, pp. 57-66.

Hayashi T.: "Application of MMCs to Engine Cylinder Blocks and Brake Disks, Chapter 6.18" in *Comprehensive Composite Materials, Vol. 6: Metal Matrix Composites* M.G. Bader, K. Kedwards and Y. Saweda, eds, Pergamon, Oxford, UK, 2000, pp. 375-379.

Chung D.D.L. and Zweben C.: "Composites for Electronic Packaging and Thermal Management, Chapter 6.38" in *Comprehensive Composite Materials, Vol. 6: Metal Matrix Composites* M.G. Bader, K. Kedwards and Y. Saweda, eds, Pergamon, Oxford, UK, 2000, pp. 701-725.

A few handbooks provide sources of engineering information and data on metal matrix composite; recent handbooks of relevance include:

The ASM Handbook 10th Edition, Volume 21 - *Composites*, D.B. Miracle and S.L. Donaldson Eds., Materials Park OH, 2001. Sections of other volumes of the ASM Handbook 10th Edition (formerly the Metals Handbook) also are relevant to metal matrix composites, *e.g.*, Volume 2 (*Properties and Selection-Non-ferrous Alloys and Special Purpose Materials*, 1990) and Volume 7 (*Powder Metal Technologies and Applications*, 1998).

Carbon and High Performance Fibres - Directory and Databook, Sixth Edition, by D.R. Lovell and T. Starr, Chapman and Hall, London, UK, 1995.

Other general references on metal matrix composites include the following monographs, edited books, or book chapters:

Clyne T.W. and Withers P.J.: *An Introduction to Metal Matrix Composites*, Cambridge University Press, Cambridge, UK, 1993.

Suresh S., Mortensen A. and Needleman A., eds: *Fundamentals of Metal Matrix Composites Butterworth-Heinemann*, Boston, MA, USA, 1993.

Everett R.K. and Arsenault R.J., ed.: *Metal Matrix Composites: Processing and Interfaces*, Academic Press, Harcourt Brace Jovanovitch, San Diego, CA, USA, 1991.

Taya M. and Arsenault R.J.: *Metal Matrix Composites: Thermomechanical Behavior*, Pergamon Press, Oxford, UK, 1989.

Chawla K.K.: *Composite Materials, Science and Engineering,* Second Edition, Springer Verlag, New York, 1998 (covers all classes of composites).

Suresh S. and Mortensen A.: *Fundamentals of Functionally Graded Materials,* The Institute of Materials, Book 698, London, UK, 1998 (scope is limited to graded MMCs).

Asthana, R.: *Solidification Processing of Reinforced Metals*, Trans-Tech Publications Ltd., Uetikon-Zurich, 1998.

Relevant review articles include:

Chawla K.K.: "Metal Matrix Composites" in *Materials Science and Technology. 13: Structure and Properties of Composites*. R.W. Cahn, P. Haasen and E.J. Kramer, eds, VCH, 1993, pp. 121-179.

Rohatgi P.K., Asthana R. and Das S.: "Solidification, Structures, and Properties of Cast Metal-Ceramic Particle Composites", *Intern. Metals Rev.*, 1986, vol. 31, pp. 115-139.

Mortensen A. and Jin I.: "Solidification Processing of Metal Matrix Composites", *Intern. Mater. Rev.*, 1992, vol. 37, pp. 101-128.

Partridge P.G. and Ward-Close C.M.: "Processing of advanced continuous fibre composites: Current practice and potential developments", *Intern. Mater. Rev.*, 1993, vol. 38, pp. 1-24.

Lloyd, D.J.: "Particle Reinforced Aluminium and Magnesium Matrix Composites", *Intern. Mater. Rev..*, 1994, vol. 39, pp. 1-23.

Hihara L.H. and Latanision R.M.: *Corrosion of Aluminum-Matrix Composites, Intern. Mater. Rev.*, 1994, vol. 39, p. 245.

Ellis M.B.D.: "Joining of Aluminium based Metal Matrix Composites", *Intern. Mater. Rev.*, 1996, vol. 41, pp. 41-58.

Tong S.C. and Ma Z.Y.: "Microstructural and Mechanical Characteristics of In-situ Metal Matrix Composites" in *Materials Science and Engineering R (Reports)*, 2000, vol. 29, pp. 49-113.

At a more introductory level, a recent source of information is the *Encyclopedia of Materials: Science and Technology*, K.H. Jürgen Buschow, Robert W. Cahn, Merton C. Flemings, Bernhard Ilschner, Edward J. Kramer and Subhash Mahajan Eds., Elsevier Science, Oxford UK (2001). Relevant entries include:

- Carbon Fibers, by L. M. Manocha
- Carbons and Graphites, Mechanical Properties of, by B. McEnaney
- Cermets and Hardmetals, by D. Mari
- Composite Materials, Microstructural Design of, by M. F. Ashby
- Composite Materials: Environmental Effects, by A. R. Bunsell
- Composite Materials: Overview, by A. Kelly and A. Mortensen
- Composites, Joining of, by F. L. Matthews

- Composites, Microstructure of: Quantitative Description, by J.-L. Chermant and M. Coster
- Composites, Physical Properties of, by P. J. Withers
- Composites: Interfaces, by T. W. Clyne
- Continuous Parallel Fiber Composites: Deformation and Strength, by L. N. McCartney and W. R. Broughton
- Continuous Parallel Fiber Composites: Fracture, by B. S. Majumdar and D. Hunston
- Cutting-tool Materials, by R. Komanduri
- Designing with Composites, by S. M. Spearing and P. A. Lagace
- Elastic properties of Composites, by A. Cervenka
- Electronic Packaging: Heat Sink Materials, by C. Zweben
- Fatigue of Particle Reinforced Materials, by N. Chawla and J. E. Allison
- Fibers with High Modulus, by A. R. Bunsell
- Fibrous Reinforcements for Composites: Overview, by K. K. Chawla
- Freeform Fabrication, by P. Calvert
- Functionally Graded Materials, by A. Neubrand
- Glass Fibers, by K. K. Chawla
- Laminates: Physical and Mechanical Behavior, by L. N. McCartney
- Mechanical Testing Methods of Fibers and Composites, by U. Ramamurty
- Metal Matrix Composites with Roughly Equiaxed Reinforcements: Microstructure and Mechanical Behavior, by D. Lloyd
- Metal Matrix Composites, Recycling of, by H. P. Degischer
- Metal Matrix Composites: Matrices and Processing, by T. W. Clyne and F. R. Jones
- Reaction Forming, by K. H. Sandhage and N. Claussen.
- Whiskers, by J. Katz

The web site of the MMC-ASSESS E.U. Thematic Network:

<http://mmc-assess.tuwien.ac.at/>

contains much relevant information on metal matrix composites, including several downloadable reviews covering various aspects of these materials (*e.g.,* heat treating, machining, joining, metallographic preparation, ...) available at:

<http://mmc-assess.tuwien.ac.at/61index.htm>.

B - Specific references of the text

Chapter 1

1. Terry B. and Jones G.: *Metal Matrix Composites, Current Developments and Future Trends in Industrial Research and Applications,* Elsevier Advanced Technology, Oxford, UK, 1990, p.105.

2. Rittner M.R.: *JOM,* 2000, vol. 52, pp. 43.

3. <http://mmc-assess.tuwien.ac.at/>

Chapter 2

1. Fitzgerald T.J., Michaud V.J., and Mortensen A.: *J. Mater. Sci.,* 1995, vol. 30, pp. 1037 - 1045.

2. Zwilsky K.M. and Grant N.J.: *Trans. AIME-Journal of Metals,* 1957, vol. 9, pp. 1197-1201.

3. McNelley T.R., G.R E., François D., McCarthy W.H., Shyne J.C., and Sherby O.D.: *Metall. Trans.,* 1972, vol. 3, pp. 1316-1318.

4. Klier E.M., Mortensen A., Cornie J.A., and Flemings M.C.: *J. Mater. Sci.,* 1991, vol. 26, pp. 2519-2526.

5. Knapp, C.K., et al.: *Composites and Method Therefor,* US Patent No. 5,477,905, 1995.

6. Skibo, M.D. & Schuster, D.M. *Process for Preparation of Composite Materials Containing Nonmetallic Particles in a Metallic Matrix, and Composite Materials Made Thereby,* U.S. Patent 4,786,467, 1988.

7. Skibo, M.D. and Schuster, D.M. *Process for Production of Metal Matrix Composites by Casting and Composite Therefrom,* U.S. Patent 4,759,995, 1988.

8. Skibo, M.D. and Schuster, D.M. *Process for Preparation of Composite Materials Containing Nonmetallic Particles in a Metallic Matrix,* U.S. Patent 4,865,806, 1989.

9. Hunt W.H.: "Particulate Reinforced MMCs, Chapter 3.26" in *Comprehensive Composite Materials, Vol. 3: Metal Matrix Composites,* T.W. Clyne, ed., Pergamon, Oxford, UK, 2000, pp. 701-715.

10. DWA Aluminum Composites: *Data Sheets provided by DWA* Chatsworth CA, USA. *<http//:www.dwa-dra.com>.*

11. Aerospace Metal Composites: *Data sheets provided by Aerospace Metal Composites (AMC)* Farnborough, UK. *(http://www.amc-mmc.co.uk).*

12. Duralcan: *Data Sheets provided by Duralcan USA, Div. Alcan Aluminum Corp., San Diego.*

13. Parvizi-Majidi A.: "Particulate Reinforced MMCs, Chapter 3.26" in *Comprehensive Composite Materials, Vol. 1: Reinforcement Materials and General Theories,* T.W. Chou, ed., Pergamon, Oxford, UK, 2000, pp. 175-198.

14. Katz J.D.: "Whiskers" in *Encyclopedia of Materials: Science and Technology,* R.W.C. K.H.J. Buschow, M.C. Flemings, B. Ilschner, E.J. Kramer and S. Mahajan, ed., Elsevier Science, Ltd, Oxford, UK, 2001, pp. 9571-9575.

15. Chawla K.K.: *Fibrous Materials,* Cambridge University Press, Cambridge, UK, 1998, p. 293.

16. Shindo A.: "Polyacrylonitrile (PAM)- based Carbon Fibers, Chapter 1.01" in *Comprehensive Composite Materials, Vol. 1: Reinforcement Materials and General Theories,* T.W. Chou, ed., Pergamon, Oxford, UK, 2000, pp. 1-33.

17. Diefendorf R.J.: "Pitch Precursor Carbon Fibers, Chapter 1.02" in *Comprehensive Composite Materials, Vol. 1: Reinforcement Materials and General Theories,* T.W. Chou, ed., Pergamon, Oxford, UK, 2000, pp. 35-83.

18. Ichikawa H. and Ishikawa T.: "Silicon Carbide Fibers (Organometallic Pyrolysis), Chapter 1.04" in *Comprehensive Composite Materials, Vol. 1: Reinforcement Materials and General Theories,* T.W. Chou, ed., Pergamon, Oxford, UK, 2000, pp. 107-145.

19. Berger M.H. and Bunsell A.R.: "Oxide Fibers, Chapter 1.05" in *Comprehensive Composite Materials, Vol. 1: Reinforcement Materials and General Theories,* T.W. Chou, ed., Pergamon, Oxford, UK, 2000, pp. 147-173.

20. Neussl E., Sahm P.R. and Flower H.M.: *Advanced Eng. Mater.*, 2000, vol. 2, pp. 587-592.

21. 3M: *Ceramic Textiles Technical Notebook,* St. Paul MN, 1998. *<http://www.mmm.com /market/industrial/ceramics/pdfs/nextel_tech_notebook.pdf>*,

22. Wilson D.M. and Visser L.R.: *Composites Part A*, 2001, vol. 32, pp. 1143-1153.

23. Maruyama B.: "Continuously Reinforced MMCs, Chapter 3.27" in *Comprehensive Composite Materials, Vol. 3: Metal Matrix Composites,* T.W. Clyne, ed., Pergamon, Oxford, UK, 2000, pp. 717-739.

24. 3M: *Aluminum Matrix Composite: Typical Properties Data Sheet*, St. Paul MN, 2001. *<http://www.3M.com/market/industrial/mmc/>*.

25. Mendelson G.: *Trackside*, 1996, vol. 7, pp. 82-88.

26. Curtin W.A.: *Advances in Applied Mechanics*, 1999, vol. 36, pp.163-253.

27. Phoenix S.L. and Beyerlein I.J.: "Statistical Strength Theory for Fibrous Composite Materials, Chapter 1.19" in *Statistical Strength Theory For Fibrous Composite Materials*, T.W. Chou, ed., Comprehensive Composite Materials, Pergamon, Oxford, UK, 2000, pp. 559-639.

28. Murakami Y., Nakao K., Shindo A., Honjo K. and Ochiai S.: in *Proc. of Int. Symp. on Composite Materials and Structures*, Proc. Conf., Beijing, June 10-13, T.T. Loo and C.T. Sun, ed., Technomic Publishing Co., Lancaster - Basel, 1986, pp. 1045-1050

29. 3M: *Continuous Fiber Aluminum Matrix Composites for High Speed Rotors*, St. Paul MN, 2000. <*http://www.3m.com/market/industrial/mmc/reinforced_rings.html*>.

30. Deve H.E. and McCullough C.: *JOM (J. of TMS)*, 1995, vol. 47, pp. 33-47.

31. McCullough C., Deve H.E., and Channel T.E.: *Mater. Sci. & Eng.*, 1994, vol. A189, pp. 147-154.

32. Fukunaga H. and Goda K.: *Bulletin of the JSME*, 1985, vol. 28, pp. 1-6.

33. Donomoto T., Tanaka A., Tatematsu Y. and Akai, T.: *Fiber Reinforced Metal Type Composite Material with High Purity Aluminum Alloy Containing Magnesium as Matrix Metal*, U.S. Patent No.4,450,207, 1984.

34. Cullough C.M., Galuska P., and Pittman S.R.: in *Design Fundamentals of High-Temperature Composites*, Proc. Conf., Anaheim, CA, USA, R.Y. Lin, Y.A. Chang, R.G. Reddy and C.T. Liu, ed., TMS, Warrendale, PA, USA, 1995, pp. 15-28.

35. Neussl E., Fettweis D., Sahm P.R., Yong S., and Flower H.M., Proc. Conf. Euromat-99, Munich, T.W. Clyne and F.Simancik Eds., VCH/Wiley, Weinheim D., 1999, pp. 119-130.

36. Wawner F.E.: "Statistical Strength Theory for Fibrous Composite Materials, Chapter 1.19" in *Comprehensive Composite Materials, Vol. 1: Reinforcement Materials and General Theories,* T.W. Chou, ed., Pergamon, Oxford, UK, 2000, pp. 85-105.

37. Miracle D.: "Statistical Strength Theory for Fibrous Composite Materials, Chapter 1.19" in *Comprehensive Composite Materials, Vol. 3: Metal Matrix Composites,* T.W. Clyne, ed., Pergamon, Oxford, UK, 2000, pp. 741-778.

38. Mortensen A., Cornie J.A., and Flemings M.C.: *Metall. Trans.*, 1988, vol. 19A, pp. 709-721.

39. Kelly A. and Macmillan N.H.: *Strong Solids,* 3, Clarendon Press, Oxford, UK, 1986, pp. 299-303.

Chapter 3

1. Michaud V.J.: "Liquid-State Processing, Chapter 1" in *Fundamentals of Metal Matrix Composites,* S. Suresh, A. Mortensen and A. Needleman, Eds., Butterworths, Boston, 1993, pp. 2-22.

2. Clyne T.W. and Withers P.J. : *An Introduction to Metal Matrix Composites*, Cambridge University Press, Cambridge, UK, 1993, pp. 318-369.

3. Mortensen A.: "Melt Infiltration of Metal Matrix Composite, Chapter 3.20" in *Comprehensive Composite Materials, Vol. 3: Metal Matrix Composites,* T.W. Clyne, ed., Pergamon, Oxford, UK, 2000, pp. 521-554.

4. Lloyd D. and Jin I.: "Melt Processed Aluminum Matrix Particle Reinforced Composites, Chapter 3.21" in *Comprehensive Composite Materials, Vol. 3: Metal Matrix Composites*, T.W. Clyne, ed., Pergamon, Oxford, UK, 2000, pp. 555-577.

5. Li B. and Lavernia E.: "Spray Forming of MMCs, Chapter 3.23" in *Comprehensive Composite Materials, Vol. 3: Metal Matrix Composites*, T.W. Clyne, ed., Pergamon, Oxford, UK, 2000, pp. 617-653.

6. Mortensen A. and Jin I.: *Intern. Mater. Rev.*, 1992, vol. 37, pp. 101-128 (1992).

7. Eustathopoulos N. and Mortensen A.: "Capillary Phenomena, Interfacial Bonding and Reactivity, Chapter 3" in *Fundamentals of Metal Matrix Composites,* S. Suresh, A. Mortensen and A. Needleman, Eds., Butterworths, Boston, 1993, pp. 42 - 58.

8. Garcia-Cordovilla C., Louis E., and Narciso J.: *Acta Mater.*, 1999, vol. 47, pp. 4461-4479.

9. Clyne T.W. and Withers P.J.: *An Introduction to Metal Matrix Composites*, Cambridge University Press, Cambridge, UK, 1993, pp. 166-217.

10. Lloyd D.J.: *Intern. Mater. Rev.*, 1994, vol. 39, pp. 1-23.

11. Rühle M.: "Stucture and Chemistry of Metal/Ceramic Interfaces" in *Fundamentals of Metal Matrix Composites,* S. Suresh, A. Mortensen and A. Needleman, Eds., Butterworths, Boston, 1993, pp. 81-108.

12. Gheorghe I., and Rack H.J.: "Powder Processing of Metal Matrix Composites, Chapter 3.25" in *Comprehensive Composite Materials, Vol. 3: Metal Matrix Composites*, T.W. Clyne, ed., Pergamon, Oxford, UK, 2000, pp. 679-700.

13. Rack H.J.: "Powder Techniques in Processing of Metal Matrix Composites" in *Metal Matrix Composites: Processing and Interfaces*, Everett R.K., and Arsenault R.J., Eds, Academic Press, Harcourt Brace Jovanovitch, San Diego, CA, USA, 1991, pp. 83-101.

14. Ghosh A.K.: "Solid State Processing, Chapter 2" in *Fundamentals of Metal Matrix Composites,* S. Suresh, A. Mortensen and A. Needleman, Eds., Butterworths, Boston, 1993, pp. 23-41.

15. Taylor N., Dunand D.C. and Mortensen A.: *Acta Metallurgica et Materialia*, 1993, vol. 41, 955-965.

16. Ward-Close C.M., Robertson J.G., and Godfrey S.P.: "Fabrication of Monofilament Reinforced Titanium, Chapter 3.24" in *Comprehensive Composite Materials, Vol. 3: Metal Matrix Composites*, T.W. Clyne, ed., Pergamon, Oxford, UK, 2000, pp. 679-700.

17. Everett R.K.: "Deposition Technologies for MMC Fabrication" in *Metal Matrix Composites: Processing and Interfaces*, Everett R.K., and Arsenault R.J., Eds, Academic Press, Harcourt Brace Jovanovitch, San Diego, CA, USA, 1991, pp. 103-119.

18. Tong S.C. and Ma Z.Y.: *Materials Science and Engineering R (Reports*), 2000, vol. 29, pp. 49-113.

19. Lewis III D.: "In Situ Reinforcement of Metal Matrix Composites" in *Metal Matrix Composites: Processing and Interfaces*, Everett R.K., and Arsenault R.J., Eds, Academic Press, Harcourt Brace Jovanovitch, San Diego, CA, USA, 1991, pp. 121-150.

20. Klier E. M., Mortensen A., Cornie J.A. and Flemings M.C., *Journal of Materials Science*, 1991, vol. 26, pp. 2519-2526.

21. Hansen N. and Barlow C.: "Microstructural Evolution in Whisker and Particle Containing Materials" in *Fundamentals of Metal Matrix Composites,* S. Suresh, A. Mortensen and A. Needleman, Eds., Butterworths, Boston, 1993, pp. 109-118.

22. Calhoun R.B. and Dunand D.C.: "Dislocations in Metal Matrix Composites, Chapter 3.02" in *Comprehensive Composite Materials, Vol. 3: Metal Matrix Composites*, T.W. Clyne, ed., Pergamon, Oxford, UK, 2000, pp. 27-59.

23. Clyne T.W. and Wither P.J.: *An Introduction to Metal Matrix Composites*, Cambridge University Press, Cambridge, UK, 1993, pp. 370-398.

24. Kouzeli M. and Mortensen A.: *Acta Materialia*, 2002, vol. 50, pp 39-51.

25. Suresh S. and Chawla K.K.: "Aging Characteristics of Reinforced Metals" in *Fundamentals of Metal Matrix Composites,* S. Suresh, A. Mortensen and A. Needleman, eds., Butterworths, Boston, 1993, pp. 119-136.

26. Merle P.: *Thermal Treatments of Age-hardenable Metal Matrix Composites, Overview on thermal treatments for metal matrix composites*, 24 pages, 2000, vol. 2. *<http://mmc-assess.tuwien.ac.at/61index.htm>*.

27. Prangnell P.B.: "Precipitation Behaviors in MMCs, Chapter 3.03" in *Comprehensive Composite Materials, Vol. 3: Metal Matrix Composites*, T.W. Clyne, ed., Pergamon, Oxford, UK, 2000, pp. 61-90.

28. Persson H.: *Machining guidelines of Al/SiC particulate MMC - Overview machining on Al/SiC MMC,* 13 pages, 2001. vol. 8. *<http://mmc-assess.tuwien.ac.at/61index.htm>*.

29. Ellis M.B.D.: *Joining of Aluminium based Metal Matrix Composites*, Intern. Mater. Rev., 1996, vol. 41, pp. 41-58.

30. Persson H.: *Guidelines for joining of metal matrix composites*, 12 pages, 2001, vol. 6. *<http://mmc-assess.tuwien.ac.at/61index.htm>*.

Chapter 4

1. Pollak P.: "Aluminum Metal Matrix Composite Activities at the Aluminum Association, Inc." in *Proc. Second International Conference on Cast Metal Matrix Composites",* D.M. Stefanescu and S. Sen, Eds., American Foundrymen's Society, Des Plaines, Ill. USA, 1994, pp. 20-28.

Chapter 5

(Note: for compatibility with the electronic database, these are listed alphabetically)

3M: company brochure, 1999.

3M: *Continuous Fiber Aluminum Matrix Composite for High Speed Rotors, report*, 2000.

Aerospace Metal Composites Limited: AMC225xe Data Sheet, 2000.

Aerospace Metal Composites Limited: AMC640xe Data Sheet, 2000.

Aerospace Metal Composites Limited: preliminary data sheet, 2000.

Alcan: Duralan datasheets, 1999.

Alloy Technology International, Inc.: product brochure, 1999.

AMETEK: HIVOL™ material datasheet, 2000.

ART, Inc.: The Mark of Excellence, 2000, vol. 7.

BDM Federal, Inc.: *Metal Matrix Composites: Sector Study*, North American Technology and Industrial Base Organization (NATIBO), 1993.

Bell J.A.E, Cushnie K.K, Warner A.E.M, Hansen G.C.: *Method of forming metal matrix fiber composites*, US patent no. 5,967,400, December 1, 1997.

Bell J.A.E., Hansen G.C.: *Continuous Carbon Fiber Nickel Aluminide Matrix Composites*, private publication.

Brown A.M., Klier E.M.: *Machineable Metal-Matrix Composite*, US patent no. 5,702,542, December 18, 1995.

Brown A.M., Klier E.M.: *Machineable Metal-Matrix Composite and Liquid Metal Infiltration Process for Making Same*, US patent no. 5,511,603, June 16, 1994.

Brush Wellman Inc.: *Beryllium Metal Matrix Composite: Avionics Materials*, product brochure MAE-002.

Brush Wellman Inc: product brochure MAE-001.

Brush Wellman Inc: personal communication, 2000.

Burke, J.T.: *Method for forming metal matrix composite bodies with a dispersion casting technique*, US patent no. 5,222,542, March 18, 1991.

Burke J.T.: *Method for forming metal matrix composite bodies with a dispersion casting technique and products produced thereby*, US patent no. 5,000,247, November 10, 1988.

Burke J.T.: *Investment casting technique for the formation of metal matrix composite bodies and products produced thereby*, US patent no.; 5,010,945, November 10, 1988.

Burke J.T.: *Investment casting technique for the formation of metal matrix composite bodies and products produced thereby*, US patent no. 5,197,528 April 29, 1991.

Ceramics Process Systems: CPS Property Summary, 1999.

Chandley G.D.: personal communications, 1999.

Chesapeake Composites: DSC™ Preliminary Data Sheet, 1999.

Donomoto T., Miura N., Funatani K., Miyake N.: SAE technical paper n° 83052, 1983.

DWA Aluminum Composites: datasheet, 2000.

Froes F.H.: *Light Metal Age*, 1999, vol. 57, p.92.

Forges de Bologne: *L'intelligence dans la transformations du metal*, product brochure.

Genma K., Tsunekawata Y., Okumiya M., Mori N., Suzuki H.: *Manufacture of metal matrix composites*, Japanese patent no. JP10306334, November 17, 1998.

Gerard D.A., Suganuma T., Mikkola P.H., Mortensen A.: "Solidification-Processed Metal Matrix Composites for the Transportation Industries" in *Proceedings of the Merton C. Flemings Symposium on Solidification and Materials Processing*, R. Abbaschian, H. Brody and A. Mortensen, eds, TMS, Warrendale PA, 2001, pp. 475-488.

Goni J., Mitxelena I., Coleto J.: "Development of low cost metal matrix composites for commercial applications" in *Materials Science and Technology*, 2000, vol. 16, pp. 743-746.

Goodfellow: Product Catalogue, 1999.

Grensig F.C., Hashiguchi D.: *Mechanical and Thermal Properties of Aluminum-Beryllium Alloy AM162*, Brush Wellman MAAB-009.

Herling D.R., Grant G.J., Hunt W.: *Advanced Materials and Processes*, July 2001, pp.37-41.

Hitchiner Manufacturing Company, Inc.: Technical Update 3D6.

Hollins M.: "Brake Discs for the Lotus Elise" in *Metal Matrix Composites VI*, The Royal Society London (mentioned briefly in the introduction to the proceedings, Materials Science and Technology,vol. 14 (9-10), 1998 but not included as an article), 1997.

KS Aluminium Technologie AG: company specification sheet.

KS Aluminium Technologie AG: *High-pressure die cast and squeeze cast engine blocks made of aluminium*, company brochure.

Lanxide Electronic Components: LEC Material Selector, 1999.

Lanxide Electronic Components: Press Release March, 1996.

Lanxide Electronic Components: Press Release June 5, 1997.

Magnesium Elektron: MELRAM Preliminary Data, 1993.

Marder J.M.: *Advanced Materials & Processes*, October 1997, pp. 37-40.

Maruyama B.: "Aluminium Metal Matrix Composites" in *Advanced Materials & Processes Technology*, Information Analysis Center (AMPTIAC) Newsletter, 1998, vol. 2.

Mendelson G.: *Trackside*, 1996, vol. 7, p.82-8.

Millenium Materials, Inc.: datasheet, 2000.

Müller-Schwelling D., Röhrle M.D.: *MTZ Motortechnische Zeitschrift 49*, 1988, vol. 2.

Parsonage T.: *Development of Aluminium Beryllium for Structural applications*, Brush Wellman MAAB-006.

PCC Advanced Forming Technology: AlSiC™ Material Data Sheet, Resource Library, 1998.

Polese: *AlSiC Heatsinks*, Product Information, 1999.

Rodriguez P. et al.: *ASME EEP-Vol.19.2 Advances in Electronic Packaging*, 1997, vol. 2, pp.1-6.

Sonuparlak B., Andrews D.J.: *Silicon Carbide Reinforced Aluminum for Performance Thermal Management Applications*, PCC Advanced Forming Technology, 1998.

Textron systems: personal communications, 2000.

White D.R., Urquhart A.W., Aghajanian M.K., Creber D.K.: *Metal matrix composites*, US patent no. 5,395,701, June 16, 1993

Warner A.E.M.: *The value of graphite in aluminium MMCs*, Inco Ltd, presentation at Second Annual Aluminium Metal Matrix Composite Meeting, 1999.

Wirth X.: "Ceramic alloys offer weight and cost gains" in *Railway Gazette International*, June 1995.

Wood J.T.: *Production and Applications of Continuously Cast, Foamed Aluminium*, Proceedings of the Fraunhofer USA Metal Foam Symposium, Stanton, DE, USA, 1997.

Wrigley A.: *American Metal Market*, November 16, 1998.

Chapter 6

1. Gerard D.A., Suganuma T., Mikkola P.H. and Mortensen A.: *Solidification Processed Metal Matrix Composites for the Transportation Industries*, Proceedings of the Merton C. Flemings Symposium on Solidification and Materials Processing, R. Abbaschian, H. Brody and A. Mortensen, eds., TMS, Warrendale PA, 2001, pp. 475-488.

2. Hunt W.H.: "Metal Matrix Composites, Chapter 6.05" in *Comprehensive Composite Materials*, M.G. Bader, K. Kedwards and Y. Saweda, eds, Pergamon, Oxford, UK, 2000 a, vol. 6, pp. 57-66.

3. Hayashi T.: "Application of MMCs to Engine Cylinder Blocks and Brake Disks, Chapter 6.18" in *Comprehensive Composite Materials*, M.G. Bader, K. Kedwards and Y. Saweda, eds, Pergamon, Oxford, UK, 2000, vol. 6, pp. 375-379.

4. Everwin P., Köhler E., Ludescher F., Münker B. and Peppinghaus D.: *Open documentation, Kolbenschmidt AG, Neckarsulm, Germany*, 1997.

5. "Quiet Revolution on the Track", *The Economist Technology Quartely*, December 8, 2001, pp. 7-8.

6. Froes, F.H.: *Light Metal Age*, 1999, vol. 57, p.92.

7. Kuylenstierna C., Storstein T.: *SAE Technical Paper*, 2000-01-2763, 2000.

8. Maruyama B.: *Advanced Materials and Processes*, June 1999, vol. 1550, pp. 47-50

9. Miracle D.: *JOM*, April 2001, vol. 53, p. 12.

10. Rawal S.: *JOM*, April 2001, vol. 53, pp. 14-17.

11. Zweben C.: *JOM*, June 1998, vol. 50, pp. 47-51.

12. Chung D.D.L. and Zweben C.: "Composites for Electronic Packaging and Thermal Management, Chapter 6.38" in *Comprehensive Composite Materials* M.G. Bader, K. Kedwards and Y. Saweda, eds, Pergamon, Oxford, UK, vol. 6, pp. 701-725.

INDEX

GPSR Compliance

The European Union's (EU) General Product Safety Regulation (GPSR) is a set of rules that requires consumer products to be safe and our obligations to ensure this.

If you have any concerns about our products, you can contact us on ProductSafety@springernature.com

In case Publisher is established outside the EU, the EU authorized representative is:

Springer Nature Customer Service Center GmbH
Europaplatz 3
69115 Heidelberg, Germany

Batch number: 10755284

Printed by Printforce, the Netherlands